Ten Technologies to Save the Planet

ENERGY OPTIONS FOR A LOW-CARBON FUTURE

CHRIS GOODALL

TEN TECHNOLOGIES TO SAVE THE PLANET

GREYSTONE BOOKS

D&M PUBLISHERS INC.

Vancouver/Toronto/Berkeley

To my father and mother

Greystone Books
An imprint of D&M Publishers Inc.
2323 Quebec Street, Suite 201
Vancouver BC Canada V5T 4S7
www.greystonebooks.com

First published in Great Britain in 2008 by *Green*Profile,
an imprint of Profile Books Ltd. (London)

Library and Archives Canada Cataloguing in Publication
Goodall, Chris
Ten technologies to save the planet / Chris Goodall. —Canadian ed.
Includes index.

ISBN 978-1-55365-525-1

1. Greenhouse gas mitigation—Technological innovations. I. Title.

TD885.5.G73G663 2010   363.738′746   C2009-906360-3

Copy editing by Iva Cheung
Cover design by Ingrid Paulson
Text design by Naomi MacDougall
Printed and bound in Canada by Friesens
Printed on acid-free, FSC-certified, 100% post-consumer paper
Distributed in the U.S. by Publishers Group West

Mixed Sources
Cert no. SW-COC-001271
© 1996 FSC
FSC

AUTHOR'S NOTE

UNITS OF ENERGY

This book uses the kilowatt to describe the power that the various technologies generate or consume. A kilowatt is the amount of electrical power necessary to light ten old-fashioned 100-watt incandescent lightbulbs, or about a third of the power used by an electric kettle.

One kilowatt of power continuing for sixty minutes is called a kilowatt-hour. The other units used in this book are megawatts (1,000 kilowatts), gigawatts (1,000 megawatts), and terawatts (1,000 gigawatts). To illustrate the scale of these figures, here are some comparisons. A typical U.S. household uses approximately ten thousand kilowatt-hours or ten megawatt-hours of electricity a year. A big fossil fuel power station generates a gigawatt or more when working, meaning that it will produce about eight or ten terawatt-hours over the course of a year. This is enough to supply about a million households. As a country, Canada consumes about 540 terawatt-hours of electricity every year and the U.S. about seven times that amount.

The book offers approximate figures for the cost of producing low-carbon energy using the new technologies that I write about in the following chapters. Unless you are an expert on the wholesale market for electric power, these figures won't necessarily mean very much to you. As a very rough comparison, I looked at the prices in some of the main electricity markets on September 17, 2009, to provide a sense of how much fossil fuel generators obtain for their output. The cost to buy wholesale electric power in September varies greatly throughout the day, usually dropping to a low point in the early morning and then rising to a peak in the afternoon when demand is at its highest. On the day I did my survey, wholesale prices ranged from $35 to about $80 a megawatt-hour in California and between about $25 and $45 in the states around New Jersey. In other words, the figures in these two places varied between about 2.5 cents and 8 cents per kilowatt-hour. Wholesale prices in Europe tend to be considerably higher than in the U.S., meaning that new forms of renewable energy may have an easier time breaking into the European electricity-generating market.

At times of peak demand or when the electricity system encounters a sudden problem, such as a malfunction in a big power station, the wholesale price of power can suddenly spike upward to a level several times higher than average, as electricity grids offer idle power stations high prices to persuade their owners to start producing electricity. At present, the wholesale price at which the owner of a power station will find it profitable to start producing electricity mainly depends on the cost of coal and natural gas, today's principal fuels for electricity generation.

I am focusing here on electricity because a future low-carbon world will probably use more electric power than we do now and less oil, gas, and coal. Using renewable sources, such as wind or solar radiation, we can generate electricity without producing large amounts of greenhouse gases, and policy-makers around the world are keen to encourage a switch away from fossil fuels and toward clean electric power.

A NOTE ABOUT GREENHOUSE GASES

This book regularly refers to carbon dioxide, the most important human-made climate-changing gas. Carbon dioxide emissions are the major part of the world's greenhouse emissions, which also include methane, nitrous oxide, and several other gases that are primarily used for refrigeration or some industrial processes.

A molecule of carbon dioxide consists of one atom of carbon and two of oxygen—hence its formula CO_2. Confusingly, we sometimes talk about the weight of greenhouse gases in terms of carbon and sometimes in terms of carbon dioxide. The crucial point is that a molecule of carbon dioxide weighs 3.667 times the weight of an atom of carbon. I have tried to be clear in the text as to whether I am referring to the weight of carbon dioxide or just carbon.

INTRODUCTION

Two or three e-mail newsletters drop into my inbox every week promising spectacular returns if I invest now in green technologies. The overexcited claims of dubious stockbrokers suggest that the battle against climate change will be won as easily as the DVD took over from video cassette. The technologies promoted in these newsletters often have a disturbing reliance on breaches of the hitherto unassailable laws of physics.

This book is more restrained. It does not claim that the world will painlessly escape from the shackles of fossil fuel dependence, quickly and cheaply building a low-carbon economy. But I hope it demonstrates that, however difficult the transition might be, the world has the tools it needs to tackle climate change. The book identifies and explores ten separate ways in which we could significantly reduce emissions or extract large volumes of carbon dioxide from the atmosphere. It also suggests that, once we have successfully switched away from coal, gas, and oil, we will find that energy costs are no higher than they are today, and perhaps considerably lower. Huge technological improvements to come will reduce the price of low-carbon energy to a fraction of what it is today. The earlier we start a systematic program of investing in new technologies that don't use fossil fuel, the sooner we will see the costs decline to the level of today's fossil fuel prices.

The following chapters steer a line between the technophiles, who believe that free markets will naturally bring about the growth of alternatives to oil, coal, and gas, and the growing number of environmental pessimists, who think that the world is hurtling toward catastrophe at increasing speed. Most of the technologies discussed are still in their infancy, and, although their prospects seem bright, none will advance rapidly without large amounts of risk capital, consistent and expensive support from governments (and therefore also from their electorates in democratic societies), and continued scientific advances. And I hope it goes without saying that these technologies are not a substitute for energy-efficiency improvements across industry and domestic life. The world needs a mix of technical advances and complementary reductions in energy use—including substantial lifestyle changes—if we are to stop and eventually reverse the rise in greenhouse gas concentrations in the atmosphere. Investment now in alternative technologies will also release us from reliance on oil and gas supplies imported from a small number of countries, not all of which bear the West much goodwill.

Some of the ten technologies in this book will fail, and it is a reasonable bet that a clear majority of the innovative companies that I briefly profile will not even exist in ten years' time. This shouldn't particularly concern us. All that matters is that those technologies that do eventually succeed are rolled out on a massive scale. Even the global warming pessimists should recognize that the world's entrepreneurs, venture capitalists, and scientists are devoting unprecedented amounts of ingenuity and hard work to the greatest challenge of our age. This is a global effort, and the following pages look at people and companies in places as diverse as Canada, China, the U.S., Ireland, Spain, Korea, Britain, India, and Australia. If the world fails to solve its climate change and energy security problems, it won't be because these individuals didn't try hard enough.

THE SECOND GLASS PROBLEM

When speaking in public, almost all specialists engaged in the climate change debate offer a positive and hopeful view of the world's ability to tackle global warming. They know that if they say that the situation is too awful and frightening, they will lose the audience's sympathy. Speakers have to be relentlessly upbeat, stressing the capacity of the world to reduce its use of fossil fuels while still improving prosperity around the globe. With a few exceptions, the public stance of climate change experts is that global warming is within our control, at least for the next few years.

Often, a reception follows the speech, and the scientist or politician speaker will stay to chat with the people who came to the talk. Glasses of wine are passed around, and the conversation moves to the actions the world needs to undertake to avert the potential of unmitigated catastrophe. I have been to many of these events, and I have noticed the same thing happen on almost every occasion. Winding down after the talk, the speaker sips the first glass and continues to say that the climate problem is within the capacity of the world to solve. But as he or she reaches for a second glass, and the alcohol starts to loosen inhibitions, the speaker begins to offer a less cheerful view. The slow pace of change in attitudes among the world's political elite is witheringly dissected. (I would use the word "glacial" to describe the rate of progress, but since some Greenland glaciers now move several miles a year, this adjective is far too generous.) The speaker notes the mounting evidence that the relatively small increases in average temperature we have already seen are having surprisingly dramatic effects. The Arctic will probably

3

have ice-free summers within a decade, major Asian rivers are likely to dry up for several months a year, biodiversity is declining at an accelerating rate, and increases in crop yields are slowing as drought, rising salinity, and increasing temperatures affect vulnerable plants. The speaker now says what he or she really believes: the world is not yet ready to make the adjustments necessary to control climate change.

Many in the climate change debate have, understandably, moved on to a metaphorical second glass of wine. They have become deeply pessimistic about society's capacity to change course quickly enough. They despairingly note that global carbon dioxide emissions appear to be rising faster than in any of the scenarios the reports of the Intergovernmental Panel on Climate Change had predicted. The recent economic recession depressed oil and gas demand, but the moment growth picks up, energy demand will race ahead. Few countries have begun the process of decoupling the growth in their economies from increasing fossil fuel use. Among policy-makers, the pessimists point out, self-delusion abounds. European governments loudly claim success in beginning to stabilize greenhouse gases, for example, but they ignore the emissions from aviation and those "embedded" in the ever-growing number of manufactured products imported from China.

It isn't that the world doesn't recognize that global warming is a problem. A large majority of people around the world agree that human activity is causing changes in the climate. And people are concerned about these changes. In an international survey conducted by the international bank HSBC in 2007, 60 percent of India's population said that climate change was one of their biggest worries. Global warming skeptics still exist in large numbers, but the majority of people, perhaps observing the increasingly obvious evidence from the natural world, accept that rapid and unpredictable climate variations are happening around them. Forests are more vulnerable to fire, storms are increasing in

4

intensity, ice packs and tundra are melting, and drought is causing starvation in water-stressed countries.

When individuals are asked whether humans can successfully
control climate change, we see large differences in the responses
between people of different countries. Those in the developing
world are much more inclined to believe that the global community can successfully stop global warming. In India, 45 percent of
respondents said we can control temperatures, but in France, the
figure was less than one in ten. The inhabitants of rich countries,
usually responsible for a disproportionate share of greenhouse
emissions, are generally not optimistic about humankind's ability
to solve the carbon dioxide problem.

So who's right? Are the attitudes that come to the surface
when sipping the second glass of wine reasonable, or are there
good grounds for the optimism widespread in India, Brazil, and
China? Is it too late, or perhaps just too technically difficult, to
reduce our economic reliance on fossil fuels?

This book argues that we have reason for very considerable
optimism. Each of the ten chapters looks at a technology or technique that could reduce CO_2 emissions by at least 10 percent of the
annual world total. All of them are comfortably within our scientific and technological reach. So, to use that ugly phrase, we should
be able to "decarbonize our economy" at an affordable price.

In fact, we can implement many of the technologies in this
book, such as zero-till farming or improved home insulation, today
with no permanent increase in costs. They will improve incomes,
make agricultural yields more reliable, or reduce household expenditure. Other technologies, including second-generation biofuels
and tidal energy, will probably be more expensive than their fossil fuel equivalents for some years or decades to come. But every
chapter concludes that with reasonably predictable technological progress we can expect our energy sources to eventually be no
more expensive than they are now. Importantly, I also try to suggest that carbon sequestration—ensuring that CO_2 is permanently

5

stored—is a readily available option, albeit at some cost. This is very good news.

Nevertheless, I don't want to suggest for one second that phasing out fossil fuels is going to be easy. After all, when I give talks on climate change, I always refuse the second drink after the speech, for fear I will let my own worries show. We should accept that some of the technologies in this book require us to make wrenching changes to the way we do things. At the moment, for example, we power our cars with gasoline or diesel. The fuel we need is available at many thousands of gas stations at the side of the roads across the world. Liquid fuels are convenient and give us unparalleled flexibility. But from a climate change point of view, moving to a world in which we all use electric vehicles powered by batteries charged with energy from renewable sources makes eminent sense.

Eventually, battery-powered electric cars will be cheaper and easier to maintain than the dinosaurs of the internal combustion era. But shifting the world's car fleet to running on electrons rather than gasoline is not a trivial task. Batteries need to improve substantially in cost, the speed with which they charge, and their capacity to store enough power to drive the car for hundreds of miles. Although the first all-electric sports cars are appearing on the road to excited reviews, this development does not mean that batteries will rapidly become the main means of automobile propulsion. We need entrepreneurs and corporations to take huge risks in moving away from gasoline. Governments must offer support and fiscal encouragement. Car users will need to tolerate the flaws of the first generation of electric vehicles. But once we have got over the humps in the road, everybody will wonder why we took so long to switch to pollution-free, easy-to-maintain, super-efficient battery cars.

Almost all of the other technologies described in this book will go through similar phases: an expensive and inconvenient introduction; a troubling period in which enthusiasm wanes and improvements appear to be slow; gradual acceptance by skeptical purchasers; and, eventually, a dawning sense that we really can do without the fossil fuel alternative.

There is one particularly easy way to attack my restrained optimism: point to the experience of the early 1970s. Then, as now, the price of oil had accelerated upward at a dizzying rate. Governments and companies around the world were eager to rid themselves of dependence on the oil cartel. Research and development (R&D) programs tried to find the best way to commercialize low-carbon technologies. Many of these R&D efforts went into precisely the same set of technical opportunities promoted in this book. The U.S. government put money into biofuels, the Chinese invested heavily in anaerobic digestion, the U.K. began research into wave power, and governments in Europe backed combined heat and power plants. They are doing exactly the same today. Unfortunately, however, they are often spending a smaller percentage of our national incomes on these technologies than they were thirty-five years ago.

In many cases, as the cynics never cease to remind us, the earlier attempts to speed the development of new energy-generating techniques were complete failures. Costs remained high, the technology immature, and consumer interest limited, despite the investment of billions of dollars of public money. Generally declining fossil fuel prices in the three decades after the oil shock of the early 1970s caused governments to lose interest, and most research efforts faded away. One of the great ironies of the last few years is that some of the scientists involved in the 1970s alternative energy drive have been brought out of retirement to restart the same R&D programs that were abruptly shut down decades ago. The people who came closest to finding an industrial-scale technology for making diesel fuel from growing and then crushing algae are back in the labs they left thirty years earlier.

Some of the other early pioneers of low-carbon technologies, perhaps conscious of the transience of government and investor

7

interest, have decided to move on. Salter's Duck, one of the first wave power–collecting devices, was designed in 1974, just after the oil shortages of the preceding year. It was a genuine advance, and its efficiency in capturing the energy in waves has scarcely been bettered since. Professor Stephen Salter, its South African–born inventor, now has the pleasure of watching wave power finally being commercialized. But in the interim, after decades of minimal official interest in renewable power, Salter has switched focus. He and his colleagues are now investigating a technology to increase low-level cloud cover over the oceans. Since low clouds block the sun's rays, Salter's scheme might help limit global warming. This initiative is one of the ten or so "geo-engineering" projects that the more pessimistic among the scientific community are investigating in the hope of dealing with—rather than trying to avert—the consequences of the carbon dioxide buildup. This book's epilogue looks skeptically at how we might use some of these schemes in a climate emergency.

Given the rapid fall in enthusiasm for alternative energy two or three decades ago, why should we believe that the current level of interest will be any more persistent? Are we simply naive for thinking that the nascent technologies of solar power, fuel cells, and advanced biofuels will ever be competitive with fossil fuels? Three forces bolster my optimism.

First, many of the technologies that looked good in 1973 but failed commercially have sustained significant price reductions since then. Many important low-carbon technologies—most obviously wind power—have come down in cost with almost predictable regularity. Consistent with the well-understood theory of the "learning curve," manufacturing costs have gone down by a similar percentage every time the number of units produced has doubled. If the experience of almost every single manufacturing industry in the world is any guide, the next doubling of the total number of wind turbines will reduce costs by approximately the same percentage. Even though the recent headlong rush into wind power caused a sharp spike in equipment prices, underlying costs will continue to fall as manufacturers gain knowledge of how to build turbines more inexpensively. By contrast, as the world pumps more oil from its increasingly depleted stocks, the price will almost certainly tend to go up. Although we continue to see far more money invested in oil exploration than into alternative energy development, the number of barrels found per million dollars spent is still declining. At some point, perhaps soon, the financial returns of investing in low-carbon technologies will exceed those from drilling for oil and gas. At that point, costs of new technologies will likely dip sharply.

Second, the world is now concerned about climate change. With the exception of a few peculiarly forward-thinking scientists, no one was worried about global warming in the early 1970s. This shift in attitude means that low-carbon technologies are more likely to garner the long-term support and subsidy that they need.

Third, leaders around the globe have an increasingly strong sense that the world is beginning to run out of minerals, or at least failing to keep up with the increase in demand. The recent simultaneous increases in the prices of metal ores, fossil fuels, and fertilizer sources such as phosphate have finally created awareness that the globe cannot be indefinitely mined. The fact that future economic growth cannot be based on unlimited supplies of raw materials, available for little more than the cost of extracting them from the ground, now seems painfully obvious. This is a sharp about-face from the attitudes of even five or ten years ago, when pessimism about the long-term availability of raw materials was confined to a few inveterate doom-mongers. Even if we did not need to reduce fossil energy consumption for climate change reasons, there are compelling reasons to find ways of living without continuous recourse to scarce and increasingly expensive materials extracted from a thin layer of the earth's crust. Low-carbon energy sources have the advantage of working with the grain of this important change in the *Zeitgeist*.

OBSTACLES TO THE TEN TECHNOLOGIES

Just because a technology is good and its financial advantages clear doesn't mean that it will be seamlessly and quickly incorporated into widespread use. In each of the ten chapters of this book, I try to note the chasms—technical and financial—that have to be crossed before we see truly widespread adoption of the most promising low-carbon opportunities. I have done this partly to rebut the accusation that this book is no more than a public relations campaign on behalf of the new industry on which I am commenting. I also suspect that we will go through several cycles of elation and disappointment before the full outlines of a low-carbon society become clear. It is better to recognize early that the road is not going to be easy.

Loss of convenience
One of the main obstacles to adopting new technologies is the ubiquity of the existing infrastructure that enables us to use fossil fuels cheaply and conveniently. Thousands of billions of dollars have been spent building natural gas pipelines and storage tanks, electricity distribution grids, huge coal- and gas-fired power stations that operate safely and reliably, mostly with few hours each year of unscheduled maintenance, and networks of oil refineries and gas stations. Switching away from the pipes, wires, buildings, and machines that have been so expensively built up over the last century or so and have served the inhabitants of prosperous countries so well won't be easy.

Some alternative energy products can fit into the existing infrastructure. Cellulosic ethanol can, for example, be mixed with conventional gasoline without requiring new cars, gas stations, or oil refineries. But other technologies require new distribution systems. Wood-based community heat and power plants, for example, rely on the installation of hot water pipes around urban areas. The large companies at the center of the fossil fuel economy—electricity generators, oil companies, and pipeline operators—have the human and financial resources to invest in projects of this scale. Few institutions have the financial capacity or skills to do the same in the low-carbon world.

Moreover, our lives are currently structured around instant and consistent access to energy. For example, the electricity system in advanced countries offers nearly universal access and reliability. Even a brief power loss in a developed nation can prompt startled front-page newspaper headlines. To expect low-carbon technologies to match this reliability and replace all the other advantages of fossil fuels within a few short years would be naive. We will go through periods when the new technologies fail, provide only intermittent supply, and cost more than their fossil fuel equivalents.

Archive films from around 1900 show many spectacular, and often very funny, failures of prototype airplanes to get off the ground. Low-carbon research will throw up similar disasters. The last few years have offered several good examples that would raise a smile were not the hopes of a hard-working entrepreneur so dashed by these events. In one case, algae grew so fast that they overwhelmed the inlet pipe at a power station where they were being tested as a way of capturing carbon. A wave power collector fell to the bottom of the ocean. A turbine blade fell off a tidal power device. Such failures provide an opportunity for skeptics to brief the press with the message that renewable technologies will never replace the electricity that fossil fuels so ubiquitously provide. If we're to tackle climate change, however, we must take a leaf out of the Wright brothers' book and not be deterred by early difficulties.

Resource shortages
Another issue facing some of the technologies profiled in this book is a shortage of key components. In 2008, wind turbines suddenly became scarce because the rapid increase in demand left manufacturers and their suppliers unable to make enough of

these surprisingly complex pieces of equipment. After decades of decline, turbine prices rose sharply. Marine energy development is being held back by a worldwide shortage of the specialized ships that can carry out operations on the seabed. At the other end of the spectrum, the growth in the number of electric cars on the road may be limited in the longer term by shortages of minerals, such as lithium, needed to make their large batteries. Similarly, some types of fuel cells, though not the ones discussed in this book, are reliant on supplies of currently eye-wateringly expensive platinum, and many solar photovoltaic companies are competing for the limited world supply of ultrapure silicon, although the shortages of 2008 turned into an enormous glut in late 2009.

As we build up the low-carbon economy, we will experience repeated bottlenecks, periods of dramatic oversupply, and painful interruptions that temporarily check progress. These problems are not a persuasive reason to hold back development of alternatives to fossil fuels—they simply mean we will experience a slower and more painful transition than we might have hoped for.

The need for scalability

The environmental movement sometimes conflates its concerns about climate change with its deeply held dislike for many aspects of the modern economy. Eco-activists often rail against impersonal and amoral multinational corporations, the gigantic and unresponsive public utilities that dominate electricity and gas supply, and the political sway of fossil fuel interests. Partly as a result, well-meaning environmentalists and green politicians often prefer to support technologies operating on a small scale. They back subsidies for solar panels on houses, small wind turbines, and wood-burning home heaters. Nothing is inherently wrong with these technologies, except that they are far more expensive for each unit of carbon saved than their full-scale equivalents. For this reason, micro-renewables will not stop climate change.

Take Germany's subsidy of small-scale solar photovoltaic installations, an unprecedentedly generous initiative. It costs the country about $7 billion in annual payments, but solar still generates less than 1 percent of the country's electricity. More importantly, perhaps, the subsidy scheme has sucked in much of the world's supply of photovoltaic panels and put them on roofs at latitudes where they are likely to generate less than half the electricity that they would have in, say, southern Spain or Mexico. Although Germany's subsidy helped build the businesses of the main Californian and Chinese photovoltaic panel manufacturers, it also substantially pushed up the price of silicon for the rest of the world for several years.

We need to apply the ten technologies in this book in ways that maximize their benefit, and this generally means large-scale implementation. Electricity from a wind turbine attached to a house well away from an ocean coast might cost five times as much as that from a properly sited large wind farm in a good location on the Atlantic seaboard. Does subsidizing the smaller turbine really make sense? Probably not. The reality—one we may not find particularly comfortable—is that the low-carbon world may have to be dominated by companies as large as today's oil and electricity companies. We need corporations that can invest tens of billions of dollars every year in huge projects in every country in the world. To fight climate change, we must use the strengths of global capitalism, not pursue an unwinnable battle of using the threat of global warming and energy insecurity to alter the way the world economy works.

As the German example shows, it's also important to put alternative energy sources in the most appropriate locations. It makes sense to focus on battery-driven cars first in countries where typical driving distances are short and gasoline is expensive. City states like Singapore and small countries like Israel are good examples. Success will be much easier to achieve there than in places where people need to drive long distances and fuel is relatively cheap. We should concentrate soil carbon improvement programs in those countries with large expanses of carbon-poor soils and well-established educational infrastructures to help farmers understand the merits of different grazing practices. Carbon capture and storage will work best in those countries with abundant deep saline aquifers. These points may seem obvious, but governments, companies, and individuals, in their zeal to appear to be doing useful things, have sometimes been distracted by the comforting acceptability of micro-initiatives rather than focusing on technologies that can be implemented at truly gigantic scales across the world.

Individual countries need to assess which technologies are most relevant to their particular circumstances and focus their limited resources on these opportunities. Cloudy Britain is wasting its money subsidizing the installation of solar hot-water units on domestic homes when it could sponsor R&D into exploiting the country's awesome resources of wave and tidal energy. By contrast, China should continue to concentrate on small-scale biogas digesters, highly forested countries like Sweden and Canada on wood-based community heat and power plants, Spain on solar energy, Denmark on maintaining its unrivaled expertise in wind power, and Australia on soil improvements. Given the importance of the car in North American society, it also seems to make sense for researchers on the continent to continue to lead the world in developing second-generation biofuels, such as cellulosic ethanol.

If individual countries focus on two or three of the ten technologies in this book, commercial success and continuing improvements in cost and usability are more likely to happen. Scattered, unreliable, and inconsistent support may actually be counterproductive, because it will divert resources from more appropriate objectives. The battle against global warming should not be a game of roulette with countries tossing a few chips toward random technologies. Research and development, public investment, and tax incentives must be thoughtfully targeted.

LINKAGES BETWEEN THE TEN TECHNOLOGIES

Suggesting that countries should focus on the technologies likely to be most appropriate for their circumstances, whether economic or geographic, but then stressing the need to understand the close relationships between each of the ten proposed solutions, may seem inconsistent. The point I am trying to make is that it will make sense for governments and companies to invest in a smaller number of promising opportunities but that countries will need to deploy the full spectrum of technologies in order to ensure an energy supply that is almost as dependable as fossil fuels are today in rich countries.

Symbiosis

Fossil fuel energy has gained its dominance partly because it is so utterly reliable. With the possible exception of wood-burning power stations and liquid fuels made from biomass, no renewable electricity technologies offer quite the same degree of consistent availability. If we are to completely run the electricity-generating systems of large countries without fossil fuel power, which is a much less far-fetched idea than most people assume, we will need to find ways of ensuring that each new technology buttresses, rather than undermines, the others.

Take wind power. The unpredictability of the winds means that individual grids, whether regional or national, cannot easily accommodate more than 10 or at most 20 percent of wind-generated power. So technologies like power storage, such as in car batteries, or by electricity sources that can be turned up and down quickly, such as fuel cells or wood-burning power stations, must complement the growth of wind turbines. If the wind stops blowing, batteries in electric cars that are plugged into the electricity system, perhaps in domestic garages, can be gradually discharged to provide backup power, or fuel cells that generate electricity for a home or business can be remotely commanded to increase their output rate.

Similarly, solar energy, which will only ever deliver direct electricity twelve hours or so a day, will need large-scale storage systems. And tidal energy, which peaks according to a predictable cycle but at varying times of day at different points on a coastline, should be widely geographically dispersed to make sure that its contribution is as consistent as possible.

One conclusion that I've drawn from my work researching this book is that energy supply in each country will probably need to be carefully planned by a central authority. The free market will be very useful in deciding which potential technical innovations offer the best opportunities, but it will probably not give us the tight integration of various complementary technologies that the world needs. This point is forcibly made when we look at the likely impact of encouraging the growth of nuclear power on the incentives to invest in low-carbon energy sources.

To an extent that policy-makers don't seem to appreciate, nuclear energy is the enemy of renewable sources of power. Nuclear plants cannot be switched on or off at short notice: to be cost-effective they must run twenty-four hours a day. Renewable sources generally also benefit from running all the time that their power source is available. Wind, solar, and marine power sources are all expensive to build and cheap to operate, so running them for as many hours as possible makes clear financial sense.

Therefore, nuclear power stations and renewable energy sources such as wind turbines are in direct competition. If a large number of nuclear plants are generating power every day of the year, other forms of supply may not be needed much. The U.K. government is talking about encouraging utilities to build new nuclear power stations that would be able to supply all the electricity the country needed at periods of minimum demand, which usually occur at about 5:30 A M. This scheme means that when demand is low, wind turbines (and all other renewable power sources) would need to be disconnected from the electricity grid, and the owners of these assets would not be paid for their electricity, a fact that makes investing in wind more risky than it would otherwise be. Unless large amounts of power can be exported to other countries freely and at reasonable prices, a large nuclear industry is incompatible with encouraging major investment in wind or any other sources of renewable electricity. Having a central energy authority may remind of us of the old days of Eastern Europe, but generators of renewable energy need to be given a clear and quite precise promise about how much other generation capacity will be constructed.

The importance of land use

We can eventually obtain most of our electricity from renewable sources directly or indirectly powered by the sun. Wind energy, wave power, and solar technologies all harness power that originated in the sun's nuclear reactions. The chapters on these technologies give figures for the percentage of the world's electricity demand that each can comfortably provide. The percentage of the earth's surface that will be needed is not large.

But we will also need to use some of the sun's energy that has been captured in plants and trees. Biomass, the technical term for energy sources created through photosynthesis, is going to be an increasingly important source of our electric power and liquid fuels. The discussion about cellulosic ethanol (see Chapter 7) shows how wood and straw can be converted cost-effectively to a gasoline substitute. But cars are prodigious users of energy. The average American car uses approximately 700 gallons of fuel a year. Even with efficient new technologies that convert wood and straw into ethanol, one such car will need the yearly cellulose output from an acre of land. If this acre has good soils for growing

17

grain, it could have produced enough wheat to feed twenty or more people for a year.

One of the great issues the world faces is how it decides to allocate land among the various competing uses. Four chapters of this book assess technologies that in one way or another use the resources provided by photosynthesis, the process of turning light into plant growth. We will need to devote land to growing woody biomass for ethanol (not to be confused with the foolishness of using foods for biofuels) as well as for the fuel in combined heat and power plants. Chapter 9 shows that we can also productively sequester carbon by digging charcoal made from wood and plant matter into arable soils, and the final chapter of the book looks at taking greenhouse gases from the air through improved techniques for pastoral agriculture and through reforestation. (Producing algae in carbon capture plants might also remove land from its alternative uses in agriculture, but because algae cultivation uses so little space, this is not a major concern.)

A titanic struggle is on the horizon. The rich world will want to use land around the globe to deliver the biomass resources to fuel its cars and generate its electricity. And it has the financial strength to achieve its wishes. Even at the historically high grain prices of the last few years, arable land is potentially worth far more growing biomass for a cellulose-to-ethanol plant than for growing wheat. At the same time, the world's people need the land to grow food. Simply put, 600 million cars are competing with the food needs of 6 billion people for the products of the land. If oil prices go up again, the increase will encourage more farmers to turn to growing biomass for conversion to ethanol, which will indirectly increase the price of other agricultural commodities.

The conflict between devoting the world's productive land for fuel and using it for food might seem to make dedicating increasingly large areas to energy crops and to woody matter that can be used as a source of biochar (see Chapter 9) impossibly difficult. Even without factoring in the impact of using land to mitigate climate change, the U.S. Department of Agriculture projects that growth rates in agricultural production will dip below the rate of world population increase. This trend will erode the substantial gains in food availability that the world has seen in the last few decades. At first sight, it looks as though we can't reconcile the need for more food production with the requirement that we devote perhaps 10 or 20 percent of usable land to producing more biomass. But perhaps this pessimistic conclusion is unwarranted. In the last two chapters, I try to suggest that just as we have mined the globe for fossil fuels and minerals, we have also mined the soil, degrading its ability to grow the food the world needs. This degradation has significantly reduced the amount of carbon held in the soil. We need to reverse this process, working to gradually improve the agricultural productivity of marginal lands. Sequestering carbon in the soil by using charcoal (Chapter 9) or by improving management of grazing land and forests (Chapter 10) means that we can make much of the land around the world much more productive, easing the conflict between the need to produce more food and the need to create more usable biomass. Improving soil health has two beneficial effects: increased agricultural yields and net carbon extraction from the air.

Some scientists have proposed elaborate machines to take existing atmospheric carbon dioxide out of the air. (I discuss Global Research Technologies' elegant solution in Chapter 8.) Like these scientists, I believe that the world needs to find low-cost solutions that directly reduce carbon dioxide levels by taking it out of the air as well as invest in technologies that give us abundant energy without adding greenhouse gases to the atmosphere. The hypothesis in this book is that improving the soil is the cheapest way of achieving net extraction of existing carbon dioxide stocks in the air. It also has the manifest fairness of delivering most of the benefit to people in low-income countries, perhaps making this proposal more politically palatable to the global community.

HOW DIFFICULT CAN IT BE?

Sometimes I hear people say that climate change is an impossibly difficult problem and that advanced societies shouldn't even bother to try to avert future warming. Instead, they claim, we should try to adapt to temperature and moisture changes as they occur. Perhaps surprisingly, those who make this claim are often among the people who usually praise the ability of modern capitalist economies to adapt flexibly and quickly to any challenges that arise. Why this group is so frightened of the temporary disruption in energy costs that a low-carbon world would endure has never been clear. Why are enthusiasts for the free market so sure that business is capable of dealing with most challenges but unable to adjust to the impact of switching from fossil fuels to renewable energy sources? This point is vitally important: widespread adoption of carbon-reducing technologies is going to be very disruptive, but the great strengths of the modern capitalist economy are almost astonishing resilience and flexibility. Free-market economies have many flaws, but they are impressively successful at finding ways around technological problems.

Let's look for a second at the scale of the task we might be setting ourselves if we act now. Advanced economies typically spend about 5 percent of their gross national product on energy—one in twenty dollars of their national income. The figure is slightly higher in the U.S. at about 7 percent. An extremely pessimistic view might be that a portfolio of carbon-reduction measures taken from the ten technologies this book features could temporarily double this percentage. This increase could conceivably persist for a few years before technical advances improved efficiencies and reduced the cost. So for five or ten years, the need to avert potentially catastrophic climate change might require the rich world to spend 10 percent, rather than 5 percent, on fuel and energy costs.

Would this increase change our societies beyond recognition? Would it impose an impossible burden on this generation or the next? Of course not. Dealing with the threat of climate change could conceivably cause a maximum cut of 5 to 7 percent in our living standards for one decade. To put that in context, the last few years have seen the greatest increase in material prosperity ever known; 5 percent is less than the growth in global GNP from 2004 to 2007. So it seems safe to say that we can accommodate all the costs of dealing with emissions reductions with relatively little disruption to our way of life. We would, in effect, be abandoning about two years' worth of economic growth. Moving from coal-fired electricity generation to wind and solar power may well be difficult, but it is not going to cause catastrophe to the modern economy, despite what the global warming skeptics say.

1

CAPTURING THE WIND

Clean power that's more reliable than you'd think

WIND TURBINES are now almost a routine sight in some parts of the world. On hills in western Spain, on Danish islands, on New Zealand's moorlands, and in the Atlantic provinces of Canada, hundreds of thousands of turbines now provide power to national electricity grids. The U.S. and China were relative latecomers to the wind business, but much of the growth in wind-generated electricity now comes from these countries. All the U.S. wind farms are on land, but developers are sizing up the coastlines for the big opportunity presented by offshore turbines. A few miles offshore, and winds are generally faster and more consistent, and the lack of ground obstructions means the flow of air is less turbulent. But offshore wind power is more expensive to develop. The turbines have to withstand twenty-five years of pounding by waves and salt spray that could corrode the electrical components. So even though the same turbine might generate 20 percent more electricity offshore than it would on a nearby hill, only brave and well-financed investors will back large wind farms in coastal waters.

Everybody is watching Cape Wind, sited in shallow water off Cape Cod, which promises to be the first offshore wind farm in the U.S. The developers struggled for nearly ten years to get the necessary permits to construct 130 large turbines that would provide enough electricity for the needs of most of the communities of Cape Cod and the islands off the southern coast of Massachusetts. When constructed, the farm will produce power that would otherwise take half a million tons of coal to generate. The proposed site is one of the best in U.S. coastal waters; wind speeds are relatively high, but the area is well protected from the worst of the Atlantic waves. The environmental consequences of installing the turbines seem likely to be relatively benign, and organizations like the Sierra Club have supported the scheme. In 2006, the Sierra Club said it "has tentatively concluded that the project does not pose a significant ecological threat to birds, marine animals, and marine habitat." Even with the support of environmental activists, the Cape Wind project has struggled against hostility from many people whose ocean views and sailing routes would be affected by the banks of turbines. Although the wind farm will be several miles offshore, a determined and effective group has fought every inch to prevent the construction of even one steel tower. Since Cape Cod was once the home to over a thousand working windmills providing mechanical power to small communities, the opposition to a field of turbines scarcely visible from the shore seems somewhat eccentric.

In late 2009, the Cape Wind organization jumped the final hurdle and was granted approval by the U.S. Department of the Interior. The business now faces the task of raising the money—an amount those spearheading the project described to me as a "ten-figure sum"—to build the farm and link it to the existing NSTAR electricity network. Raising money for projects of this size and unusualness is far from easy in today's risk-averse financial markets. Banks and investors have to take a gamble—will future electricity prices be high enough to repay the capital? One particular advantage of the Cape Wind location is that the off-shore breezes characteristic of the hot summer afternoons will turn the turbines at times when electricity is at its most valuable. Mark Rodgers, Cape Wind's communications director, told me that he felt confident that the project would be financed by the end of 2010 and would ship its first electricity two or three years later. This huge scheme demonstrates both the importance of wind power and the serious obstacles that it faces in many parts of the world.

Adam Twine tends an organic farm not far from Oxford in southern England. The surrounding areas are flat and low lying, but Twine's land occupies a small and windswept plateau. Down below, in the far distance, the cooling towers of Didcot power station are the most visible marks on the landscape. Didcot is a decaying coal-fired generator due to close in a few years because it cannot meet the latest European emissions regulations. Twine's fields are not ideal for wind power—central England has far lower speeds than the western coasts and many other regions around the world—but he decided in the mid-1990s that he wanted to build a wind farm owned by the local community, sited as a perfect contrast to Didcot, the single largest source of carbon dioxide in the prosperous southern heartland of England.

As with the Cape Wind project, the struggle to get the turbines constructed was a long one. Just getting planning permission took the better part of a decade. Although Didcot's six huge cooling towers and the multiple power lines trailing away from the station have already had a huge impact on the landscape, local resistance to the visual effect of the turbines was fierce. When Twine finally obtained approval, a protracted process of fundraising began. By the time the capital was raised, a worldwide shortage of components had pushed the prices of turbines up 30 percent, so more cash was needed. With a few grumbles and support from Britain's Cooperative Bank, the shareholders obliged, raising the final installment with a few days to spare in spring 2007.

In February 2008, the wind farm started producing electricity. Five 1.3-megawatt turbines now rotate sedately (and very quietly) whenever the wind blows. Over two thousand people own shares in the development. Some invested because of a passionate belief in renewable energy; others because the venture promised good financial returns. So far, Adam Twine says that the output from the wind farm has more than delivered on the promises made, and its investors have already received their first dividend payment.

As Twine's farm shows, new wind farms are already good investments in many parts of the world. The best returns come from buying the largest possible turbines, all from a single manufacturer, and installing as many as possible in the local area. This approach reduces the costs of connecting the wind farm to the electricity grid and minimizes the amount paid for yearly maintenance. In countries such as Portugal, the largest wind developments are now obviously competitive with fossil fuel sources of electricity. BT, the U.K.'s largest telecommunications company and the user of more than half a percent of the country's electricity, says that its wind turbine construction program, planned to provide a quarter of its needs, is easily justified to its shareholders as making good financial sense.

The years 2006–2008 saw a sharp rise in the price of turbines as the steel for the supporting column and copper for the turbine wiring suddenly cost far more than ever before. The rapid growth in demand for turbines also caused production bottlenecks for some of the eight thousand components in a typical turbine. The shortages have now eased and costs have fallen, a downward trend predicted to continue, with expected costs falling from about $1,200 per kilowatt of generating capacity down to perhaps $800 by 2013—roughly equivalent to the capital cost of a new gas-fired power station. The full cost of Adam Twine's wind farm came to almost twice today's average, inflated by the relatively small size of the development and the expensive struggle to get permission to build it. Cape Wind—because it is offshore—will also be far more expensive to install.

As the critics of wind power never tire of pointing out, turbines do not generate their maximum power all the time. They only produce their full output when the wind is blowing strongly. But not *too* strongly: above a certain wind speed, the machines shut down to prevent the blades from rotating too fast and damaging the turbine. Averaged across the year, a 2-megawatt turbine in a reasonable location will typically produce only about a third of this figure—about two-thirds of a megawatt. The wind farm on Twine's land will probably generate about 13 gigawatt-hours in its first year. This figure sounds impressive, but the old dinosaur of a coal-fired power station down the road at Didcot will produce the same amount of electricity in a busy afternoon. It would take nearly a thousand wind farms the size of Adam Twine's to replace just one power station of Didcot's size. The huge Cape Wind field will only replace about 10 percent of the output of one of the largest U.S. coal-fired plants.

Given their relatively small output and inconsistent performance, are wind turbines a genuinely useful tool in the fight against climate change? The answer to that question is an emphatic yes, and this chapter explains why.

THE POWER OF THE WIND

Wind arises from variations in atmospheric pressure between different parts of the world. Air tends to flow from high- to low-pressure areas, with the speed of the wind depending on the gradient between the wind cells. The ultimate cause of these pressure differences is the differential amounts of solar heating across the globe. We can therefore think of wind as an indirect form of solar energy. A small fraction of 1 percent of the light and heat energy that the earth receives gets turned into the moving, or "kinetic," energy of the wind. We can capture this energy using windmills or wind turbines that slow down the speed of the air, transferring power to the rotation of the blades.

A wind turbine can be thought of as the opposite of an electric fan. A fan uses an electric motor to turn the blades when the electricity is turned on; a turbine does the reverse. The rotating arms turn gears, which then rapidly rotate an electrical conductor, usually a dense mesh of copper wire, inside a powerful magnetic field, inducing electricity to flow.

A wind turbine cannot capture the full power of the wind. The theoretical limit is just under 60 percent of the energy in the flow of air. And the amount of electricity generated by the rotation of the blades is only equivalent to about 70 percent of the energy captured, even in an efficient new turbine. Even with these disadvantages, wind is still a very productive source of electric power, comparing favorably with solar photovoltaic panels, which turn less than a fifth of the energy they receive into electricity.

The secret of wind's success is the sheer mass of moving air that passes through the rotating blades of a turbine. Air may seem almost weightless to humans, but each cubic yard actually weighs almost two pounds. A strong gust consists of air moving at perhaps 40 miles an hour, or 55 feet per second. This means that every second, over 37 pounds of air pass through each vertical square yard. This motion contains a substantial amount of energy, with the power in the wind proportional to the cube of the speed of the air. In other words, a wind turbine in a 14-mile-per-hour air flow will generate almost 60 percent more power than one in a breeze of 12 miles per hour. (This is why it is so important to choose windy sites for turbine locations.) At 14 miles per hour, which is little more than a gentle breeze, the motion of the wind contains about 18 watts of power per square foot. This is less than the full power of the midday tropical sun, which delivers more than a thousand watts in the same area, but wind is easier to convert to electricity and will often blow for the full twenty-four hours in the day, not just during daylight hours.

Of course, the amount of power that a wind turbine can capture is also linked to the area of the circle swept by its blades. The very biggest new turbines have arms that are 200 feet long: in a 40 mile-per-hour wind, about 200 tons of air will pass through the blades' circle every second, with a usable energy of more than 2 million watts. By comparison, a tiny domestic wind turbine with blades just over 3 feet long covers a little more than 30 square feet, capturing up to 600 watts. Somewhat counterintuitively, the large turbine doesn't sweep sixty times the area of the domestic turbine; it covers over three thousand times as much.

Wind turbines will probably stop increasing in size soon. There's talk of giant 7-megawatt turbines for offshore installations, but the limit may be 5 megawatts. The problem is that longer turbine arms, while providing more power, are also subjected to more stress. As an arm swings downward, it stretches under its own weight; as it swings upward, it becomes fractionally compressed. Repeated millions of times a week, this stress will destroy all but the strongest and most flexible materials— and the longer and heavier the blades, the greater the forces they need to withstand.

WIND'S GROWING IMPORTANCE

Only about 1 percent of world electricity demand today is met by wind, but the figure varies enormously around the world. Some areas of Germany generate more wind energy than their total power needs. Almost 20 percent of Danish electricity comes from wind, and the figure is similar in Prince Edward Island. The local electricity companies can accommodate these high levels of wind power because they have the freedom to export excess power when the wind is blowing hard and import electricity when the air is calm. Denmark's access to Norwegian hydroelectric power is particularly important.

The U.S. and Spain are adding the largest amounts of new generating capacity every year. Wind energy in India and China is also becoming increasingly important: in China, the amount of wind generation has doubled every year over the last three years. By 2015, China may have 50 gigawatts of wind capacity, or about half today's global total. In developing countries without a national electricity grid, wind power combined with large batteries will often represent the cheapest reasonably reliable way of generating power for small communities.

There's no shortage of windy sites left to exploit. One study put the average power in the global winds at any one moment as about 72 terawatts—around thirty times the world's electricity requirements, or ten thousand times the wind power we currently generate. And this estimate only includes sites with average wind speeds above 15 miles per hour, a level usually only met at coastlines or on the tops of hills. No one pretends that we can capture this entire potential, but we will be able to use wind to provide a good fraction of total world energy needs, and we can expect the rapid growth of the industry to continue for several decades.

Some of wind's growth is being pushed by subsidy schemes. Spain's 30-percent annual increase in wind power is propelled by price guarantees for the electricity that the turbines generate. But most experts now think that onshore wind turbines are close to competitive with traditional forms of electricity generation, at least in windy locations. Fairly assessing whether wind is cheaper or more expensive than gas turbines or coal-fired stations is surprisingly difficult. The assessment critically depends on assumptions about inflation, interest rates, and how long the turbines will last. And, of course, it depends on the price of fossil fuels. Nevertheless, the trend is unambiguous: wind is going to become a relatively inexpensive provider of power, and if fossil fuels continue to increase in price, this advantage will become more pronounced. Wind generation has its problems and complexities, some of which are discussed later in this chapter, but it provides us with the best possible example that

30

technological progress, heavy investment, and government help can push a new technology forward. The cost of wind power has probably fallen by a factor of ten in the last twenty-five years, and we can reasonably hope that some of the other infant technologies in this book will improve to a similar degree.

Wind provides a little less than 4 percent of the European Union's electricity today, four times the average for the world as a whole. The trade body for European wind thinks that this figure will rise to about 13 percent in 2020 and continue to increase rapidly thereafter. This increase would mean installing around 10 gigawatts of wind capacity each year over the next decade or so, which equates to thousands of new turbines annually, but since the new capacity installed in 2008 alone was almost 9 gigawatts, the target seems to be well within reach.

Getting to this level will require capital expenditure of almost $15 billion per year, even if turbine prices fall as expected. For this money, AREVA, the main European nuclear construction company, says that the continent could have two or three atomic power plants, although the final pages of this book cast some doubt on whether nuclear power can be delivered at this price. Three nuclear stations would have a capacity of almost 5 gigawatts, and typically, they would operate at full power more than 90 percent of the time. Even in windy offshore locations, wind turbines with a total maximum power of 10 gigawatts will provide at most 40 percent of their rated power, or just 4 gigawatts. So the math is quite simple: capital investment of $15 billion a year would buy Europe more low-carbon energy if invested in nuclear power than in wind.

However, once the infrastructure is constructed, wind energy is close to free—the cost of annual maintenance is usually a small percentage of the value of the electricity generated. Nuclear fuel is not very expensive, but nevertheless, its costs help equalize the price of the two forms of electricity generation. Add in the unknown costs of indefinite safe storage of nuclear waste, and wind seems only a little more expensive than nuclear energy. It will eventually be cheaper, particularly for turbines on windy coastlines.

So why do the big power companies in many countries still hanker after more nuclear plants? The reason is probably that nuclear plants are large and centralized, and they work around the clock. A big corporation can manage a small fleet of nuclear plants far more easily than it can control a network of thousands, perhaps tens of thousands, of turbines spread across large numbers of sites. The nuclear option also simplifies matching electricity supply with customer demand; if the company relies on erratic wind supplies, it will frequently be forced to buy power from alternative sources at unpredictable prices. In other words, nuclear generation works well for the big companies that dominate power generation in most countries, even though it will probably not deliver lower costs than power generated from large land-based wind farms.

Low and predictable running costs also help wind compare well with fossil fuels. Once the turbine is placed on top of its tower, virtually free electricity will be generated for the next twenty-five years or so. By contrast, the world now assumes that coal, gas, and oil are going to get increasingly expensive. So investing in wind mitigates the burden from increasing prices of other fuels. Wind has an additional advantage, too. Because its fuel is free, the turbine owners will generally always be able to sell their electricity at a profit. By contrast, the main fuels for power stations—gas and coal—can swiftly vary in price in relation to each other. The hundreds of millions of dollars invested in a coal-fired generator may produce nothing for months if the price of coal rises too high compared with the cost of gas. Didcot coal-fired power station, a few miles from Adam Twine's wind farm, sat idle over much of the winter of 2008 because an unpredicted spike in the price of coal meant that it was uneconomical to run the plant.

Reliance on fossil fuels has a real cost to the economy if consumers and manufacturers can't guess what energy prices are likely to be months, years, or decades in advance. One of wind's primary but often underestimated virtues is that it delivers electricity without such financial volatility. The output of a wind farm may be uncertain, but the cost is not. And, of course, wind power is independent of political intervention—countries that invest in wind are less reliant on the two or three countries that provide much of the world's natural gas.

KICK-STARTING AN INDUSTRY

Denmark began to build substantial numbers of wind turbines in the 1990s and became the first nation to generate a significant fraction of its electricity from this source. The early years were characterized by the installation of hundreds of what are now considered very small turbines. The owners were local co-operatives and farmers, and these pioneers allowed Denmark to develop a world-leading wind-turbine industry.

In recent years, developers have replaced these small entrepreneurial groups, constructing much bigger farms using more powerful turbines. But the community approach is still important. Take Middelgrunden, a huge wind farm in very shallow water two miles outside Copenhagen Harbour. Consisting of twenty 2-megawatt turbines arranged in an elegant ellipse, the farm is half-owned by an electricity utility and half-owned by over eight thousand individual shareholders, making it the largest mutually owned wind farm in the world. The farm was constructed in 2000, contributes about 3 percent of the electricity needs of Copenhagen, and offers its individual investors competitive financial returns. According to one study, Middelgrunden delivers power for less than 5 euro cents (about 7.5 U.S. cents) per kilowatt-hour, which is certainly no more expensive than electricity generated from gas.

33

Denmark and its neighbor Germany have both demonstrated the usefulness of developing community support for wind turbines by encouraging small investors to participate in the investment and then earning financial returns. In places like Portugal, commercial developers have also provided funding for improved public facilities for the local area. Surprisingly, this model has been slow to catch on elsewhere. Although individual investors or public authorities demand lower financial returns than big companies, the growth of mutually owned wind has been slow outside northern Europe.

Denmark's early support for wind has had several important repercussions. Among other benefits, the country became the world leader in the manufacture of turbines. Two of the world's largest manufacturers are based in the country, and Suzlon, a company based in India, runs its international marketing from Aarhus in the Jutland Peninsula. The other large manufacturers are from Spain, Germany, and the U.S., all countries that have successfully encouraged the growth of wind power, confirming the connection between the expensive process of backing a new technology and the benefit of building a successful industry that can then export its products to the rest of the world. Countries thinking about investing in other nascent low-carbon technologies should bear this point in mind. Denmark's support for wind has helped create and sustain a highly valuable manufacturing industry, employing hundreds of thousands of people in good jobs. Eventually, manufacturing leadership will pass to countries with lower labor costs than Denmark, but the country will still capture many of the benefits from its pioneering role.

34

Suzlon, one of the fastest-growing participants in the industry, already manufactures its turbines in India, but almost all of its very substantial research and development work is carried out in Europe. Its lower manufacturing cost base will help it grow its world market share from today's level of about 9 percent. In its last financial year, its sales almost doubled compared with a global increase in turbine sales of no more than about 40 percent. Suzlon is particularly well placed to supply China, with its almost insatiable need for power and its wealth of windy locations, such as Inner Mongolia.

MICROWIND

Even before factoring carbon dioxide into the equation, big wind farms seem competitive across a wide range of locations. For a home a long way from the electricity distribution network, a turbine on a tower in the garden may provide electricity at a lower price than a diesel generator, especially if oil prices continue to rise. However, for homes and businesses already on the electricity grid, small-scale wind generation will generally not be as financially attractive.

The best way of demonstrating the economics of microwind is to compare the cost per kilowatt of generating capacity. Now that the shortage of large turbines has ended and as further technical progress is taking place, a large wind turbine will probably cost a total of less than $1,000 per kilowatt of maximum power. For comparison, a very small domestic turbine may cost as much as $6,000 for a kilowatt of peak power—six times as much. As importantly, today's commercial turbines are elevated many dozens of feet above the ground. At these heights, the wind speed is far greater than near the surface. For example, the U.K. wind speed database says that the average wind speed where I live in the city of Oxford is 10 miles per hour at roof height but 13 miles per hour at 150 feet above the ground. The winds would be even higher at 250 feet, the typical elevation of a new commercial turbine. Because the power in the wind is proportional to the cube of the speed, the amount of energy available to collect with a 150-foot mast is over twice what a small rooftop turbine would achieve near my home. Lastly, a tall turbine tower in a commercial wind farm will likely be on a site with little wind turbulence, enabling the turbine to capture more wind power *and* reducing the stresses on the equipment and therefore its need for maintenance.

All in all, for $1,000 invested in a domestic wind turbine, the owner will get less than one-tenth of the electricity generated from an investment of the same amount of money in a large wind farm in a windy part of the world. This is not to say that small-scale wind is bad but merely that for maximum climate change impact, we need to put as much money as possible into turbines that can generate megawatts, not kilowatts.

Not everyone shares this view. Advocates of microgeneration claim, quite correctly, that small wind turbines have the advantage of being visible symbols of a household's commitment to low-carbon electricity. And although the first generation of domestic wind turbines have been widely criticized for not producing as much electricity as their vendors claimed, supporters of microgeneration say that advances in price and efficiency are possible in small wind turbines as much as for their larger cousins.

Also true is that some of the problems associated with turbulent air around buildings and trees may be avoided with wind turbines that do not use the classic three-bladed propeller design. Instead, manufacturers will probably focus in the future on producing what are called "vertical axis" turbines, such as the ones produced by Mariah Power in Reno, Nevada. Such small microturbines will probably be cheaper to build, install, and maintain than conventional three-bladed models of the same size.

36   The Mariah Power wind turbine costs about $9,000, including installation. Its manufacturers say that in a location with average wind speeds of 11 miles an hour, which is breezy but not exceptionally so, it will generate 1,800 kilowatt-hours a year—about a quarter of the electricity needs of a typical North American home. The cost of the machine is a major improvement on existing models, but even with generous government rebates,

it will still take well over a decade before it pays back its owner's investment, compared with as little as four years for a large-scale commercial wind farm.

Technological improvements notwithstanding, cooperatively owned wind farms remain a more exciting proposition than microwind. They encourage support for renewable energy across communities while offering far greater energy output and financial returns. One developer, Goodhue Wind in Minnesota, is developing a large 78-megawatt wind farm with the active financial participation of people in the local area. Many communities around the world could develop a similar strategy of revitalizing their district, lowering the cost of power, and providing some good "green" jobs.

OFFSHORE

At the other end of the range of turbine sizes, what about the viability of wind farms placed offshore? In some countries, but by no means all, onshore wind is unpopular on the grounds of appearance, which is tending to slow the rate of development. And even in areas that accept, or even actively admire, wind turbines, we will eventually run out of good sites. These difficulties may be prevented in the future by moving new wind turbines out to sea, where aesthetic objections are more limited and ecological damage appears to be less severe. Bats, for example, which are threatened by onshore turbines, are unlikely to fly miles out to sea to find food. So it seems easy to suggest that moving wind generation offshore makes good sense. In April 2009, U.S. Interior Secretary Ken Salazar committed the U.S. government to active development of offshore wind, saying that the coastal resources of almost 2,000 gigawatts exceeded the entire electricity demand of the lower forty-eight states. The U.K. government has publicly suggested that 33 gigawatts of offshore turbines might be built off its coasts in the next decade or two, enough to provide almost a fifth of all the country's electricity. Other countries have announced similarly ambitious targets.

However, offshore wind power is in its infancy. Although its potential is vast, the engineering challenges are far more substantial than building even the most remote wind turbines on land. Several significant projects have been completed around the world, but most developers are hesitating before committing to major investments.

The London Array is perhaps the most ambitious plan for a large-scale offshore farm. Intended to eventually include over 340 wind turbines, averaging about 3 megawatts in size, the farm will sit more than 12 miles from the Kent coast in southern England in shallow waters at the end of the Thames estuary. Once completed, the array will produce enough electricity at peak output to power a quarter of the homes in London. A reasonable estimate of the impact of the London Array on the U.K.'s emissions is a reduction of 1.5 to 2 million tons of carbon dioxide per year.

The array will probably cost over $5 billion to build, and even this figure is constantly edging upward, pushed by the high price of steel and other construction materials. A further problem inflating its costs has been the worldwide shortage of vessels to carry out complicated offshore installations. Another complication is that only two or three manufacturers currently build turbines that can survive years of storms in salty water. Even as the price of turbines reverts to normal levels, offshore-ready turbines are always going to be more expensive than their land-based equivalents.

Because the turbines are offshore—where winds are stronger and more reliable—electricity output per unit of generating capacity will be greater at the London Array than at a typical land-based wind farm. The Cape Wind array of turbines off the Massachusetts coastline should produce even more. The turbines will be rotating a larger fraction of the time. But offshore wind developers need to invest four or five times the amount that they would for an onshore wind farm, for each unit of generating capacity. So although offshore turbines produce more electricity, this advantage is outweighed by the huge cost of installing extremely robust turbines in deep water.

As the industry grows, this incremental cost will erode. Extra offshore construction vessels will be built, more suppliers will enter the market for rugged turbines, and installation techniques will improve. However, offshore turbines are always likely to be significantly more expensive than their land-based cousins, and building the foundations for the turbine towers will remain a much more difficult task than completing the groundworks onshore. Whether offshore wind power, even coming from huge farms of hundreds of 5-megawatt turbines on 300-foot towers, will ever become cheaper to produce than electricity from conventional power stations is an open question.

Some wind developers are very aware of this problem and have begun to work on designing turbines that float on rafts anchored to the seabed. This approach will allow the developers to avoid the huge costs of constructing underwater foundations and will also mean that the windmills can be installed much farther out to sea so that they are completely invisible from the shore. These turbines will never be as powerful as the biggest seabed-mounted models, but the lower installation cost may well outweigh this disadvantage. Statoil, the Norwegian oil company, proposes to place a 2.3-megawatt turbine on a platform similar to an offshore oil-loading buoy 6 miles from the Norwegian coast. The company stresses this installation is an early trial, but if it works, it may help significantly reduce the costs of offshore wind power.

One might think that offshore wind farms escape most of the planning permission issues that impede onshore construction in many parts of the world. However, getting the planning permissions and agreements to connect the London Array to the national electricity grid has been complex and unrewarding. The onshore substation that will receive the electricity as it arrives on land caused a particularly fierce battle. Although the coastline in the part of Kent that will host the substation is not of any great beauty, the struggle to obtain the necessary permits to construct the building and its outdoor apparatus was protracted to well over a year by inquiries and consultations. Local residents were particularly concerned by the prospect of large numbers of heavy trucks passing an elementary school on their way to the site during the construction period. The opponents of the substation said there would be thirty trucks a day using local roads; the developer of the array insisted that the figure was two. This apparently minor debate was one of the two or three issues that significantly delayed the U.K.'s most important new source of renewable energy. Similar arguments were heard over the landing point of the Cape Wind electricity at Barnstaple on Cape Cod. These examples illustrate the scale of the planning problem faced by the wind industry in North America and elsewhere: the London Array, a project that will save up to 2 million tons of carbon dioxide emissions, worth about $30 million per year in today's carbon market, can be delayed for years by issues such as the routing of a relatively small number of heavy trucks. To many people, clearly, the climate change threat does not appear sufficiently imminent or severe to offset any adverse effects on their local community.

UNRELIABILITY—REAL AND IMAGINED

Large onshore and offshore wind farms are going to provide increasing amounts of power over the next decades. Wind may eventually provide 20 or 25 percent of the total electricity requirements of many large countries. Nevertheless, wind has its problems. Although some people delight in the smooth elegance of rows of three-bladed turbines on the horizon, others think that they despoil the landscape. These opponents usually first attack wind power for its perceived ugliness, but in the next breath they also berate it for "intermittency." The opponents' key question is this: how can a source of electricity that is so unreliable really be worth investing in?

First, we need to be clear that wind is only really unreliable in one particular way. If the wind is blowing hard now, it will probably also be doing so in a few minutes' time. Very little unreliability there. And, year after year, wind turbines will produce approximately the same amount of electricity over a twelve-month period. We have good years and bad years for wind, but annual electricity output from a turbine will stay within well-understood bounds. In that respect, wind turbines are at least as reliable as an old coal-fired or nuclear station, where output can vary enormously because of maintenance needs or equipment failure. Wind is indeed "intermittent," then, but only in the sense that some weeks it blows hard and in other weeks it doesn't. And even these variations are more predictable than you might think.

So out of the three types of intermittency, wind power only really matches one of these definitions—week-to-week unpredictability but not minute-to-minute or year-to-year. But let's look first of all at minute-to-minute variability. From the viewpoint of the operators of the electricity grid, very short-term reliability matters most. If the total output from all wind turbines on the system did suddenly drop, it would be a significant problem, because a well-functioning grid must match electricity supply and demand very precisely. But the wind speed at a single location in the next few hours is actually reasonably predictable. And, perhaps more importantly, changes in wind speed tend to be smooth. The wind very rarely drops erratically, unpredictably, or quickly, although this could, in theory, happen.

Wind's short-term predictability means that this power source is much more easily accommodated in a nationwide electricity grid than might be first imagined. Despite what anti–wind power campaigners sometimes claim, the people who run our electricity systems do *not* need to keep an equal amount of coal-fired power generation ticking just in case the wind suddenly drops. They need a small reserve (perhaps 15 percent of the total wind-generating capacity) available at short notice, but this is little more than they have now anyway. Any national electricity distribution system already has to have power stations ready to start generating electricity at very short notice in case a large power station suddenly fails. Moderate amounts of wind, perhaps up to a sixth of total electricity production, can usually be managed without any changes in the complex way that electricity trading and distribution systems work.

As the total number of wind turbines increases, short-term variability actually becomes easier to handle. Typically, the turbines will be spread over a wider area—perhaps the whole country—and when the wind is quiet in one place, it is likely to be blowing strongly in another. The total electricity output from a thousand turbines varies far less than the power generation from ten.

But a grid can only easily accommodate a limited amount of wind power. In some short periods during March and April 2008, wind turbines in Spain were generating up to 40 percent of the country's electricity requirement. At this peak, the grid operator required electricity generators to disconnect some turbines from the network. Although the wind levels were extremely unlikely to drop unexpectedly, even small percentage changes in wind output might have overwhelmed the country's limited ability to import power instantaneously from France or switch on backup power stations. This is a genuine concern arising from the growth of wind power—the tiny but important risk of unpredicted short-term variability when wind is responsible for a very large percentage of electricity generation.

The solutions to this problem are reasonably simple, although not necessarily cheap. We could build inexpensive gas-fired power stations that can be very quickly fired up in the event of unexpected shortages of supply—though this approach means cost and carbon emissions. Alternatively, we could try different approaches that do not require burning more fossil fuels whenever wind speeds don't match expectations. We have three main routes for achieving this. First, we can make importing power from remote locations easier. Second, we can store electricity. Third, we can introduce systems to manage electricity demand at short notice so that it matches the available supply.

Every country in the world that relies on increasing amounts of wind, marine, or solar power will probably need to use all three of these mechanisms to align short-term supply and demand. In the U.S., this three-pronged approach is appropriately called the "smart grid." The construction and operation of this new kind of grid are fascinating challenges to engineers and also to the mathematicians who will use statistical modeling to minimize the risk of not having enough power or, perhaps even more expensively, having grossly excessive power production for many hours a week.

Elsewhere, the standard approach, which we might call the "twentieth-century model," simply tries to predict changes in demand and then adjusts supply to meet these variations. For example, an advertising break in a popular television program produces a sudden surge in demand as lights are turned on and kettles are boiled. Just before this happens, a large power station needs to be warmed up so that it is ready to start producing electricity the moment demand begins to surge. Supply simply aims to match moment-to-moment demand. This model is both costly and carbon intensive, because power stations have to be held in reserve, burning large amounts of fuel even when they are not supplying power to the grid.

The smart grid is more efficient—and it's also compatible with the incorporation of large amounts of power from wind and other unreliable energy sources. Let's look in a little more detail at its three main approaches.

Importing remote power
If the wind suddenly drops on the Atlantic coast of Spain, it is statistically extremely improbable that Denmark will suffer at the same time. So if the electricity grid connected Spain in

43

southern Europe and Denmark, far to the north, spare power could flow southward at very short notice. At the moment, most countries have poor connections to their neighbors. In North America, the much larger distances have magnified this problem, with transmission towers able to move only limited amounts of electricity from one region to another. When Spanish wind generation peaked in spring 2008, the link between France and its southern neighbor was not robust enough to handle the possible demand from Spain if the wind suddenly dropped. It can currently handle only about 5 percent of Spain's total demand. Growth in wind capacity around the world must be accompanied by major investments in power distribution networks to increase the number and size of the electricity transmission links between different countries and between regions inside countries. The aim is to be able to move electricity nearly instantaneously from countries in surplus to those in deficit. These new high-voltage links will need to be paid for, but most independent studies show that the costs are unlikely to add more than a fraction of a cent to the price of power. Building bigger, better, and more robust electricity grids simply must happen if we are to significantly increase renewable power use. Most governments understand this reality, but many still do not.

Storing electricity to meet short-term needs
We can store electricity in batteries, but this approach is expensive. It does not provide a large reserve and cannot yet be used to back up wind turbines. Today, the cheapest way of providing a reserve of usable power is through a system called "pumped storage." When electricity is abundant or cheap, it is used to move very large quantities of water uphill into a storage reservoir. When demand is very high, or when the grid needs power urgently, the water in the high reservoir is released, turning turbines as it falls back into the lower reservoir. A good pumped storage system can start generating large amounts of power as

little as fifteen seconds after a request from the electricity grid. The U.K.'s largest storage reservoir, in North Wales, can supply as much power as the biggest power station in the country and is invaluable on the infrequent occasions when a large power plant does fail without warning. When the higher reservoir is full, it can keep generating for several hours before it runs out of water, giving the people who run the electricity network enough time to start other power stations.

For many countries, pumped storage is the best-established way of dealing with immediate needs for power. The Spanish electricity grid now has about 3 gigawatts of pumped storage. As in the U.K., this capacity was built to help insulate the power grid from the effects of spikes in demand or major power station failures. But now, in addition, the country has over 15 gigawatts of wind power, expected to rise to more than 20 gigawatts by 2010. The ratio between the current storage capacity and the possible fluctuations of wind output is not great enough. As a larger and larger percentage of electricity comes from renewable sources, the need for countries like Spain to build water storage reservoirs will grow. This task is not necessarily easy to achieve. A satisfactory site must accommodate two large reservoirs, not far apart, one much higher than the other. Many countries will have relatively few locations that meet these criteria. Moreover, suitable sites will tend to be in hilly or mountainous places far away from the main electricity transmission tower routes, making connecting the reservoirs to the distribution grid difficult.

One alternative is to establish lagoons at sea—large areas of ocean surrounded by a high wall. When electricity is plentiful, pumps bring seawater into the lagoon, creating a gradient. If the wind falls unexpectedly, this gradient can be used to drive turbines for near-immediate electricity. These barrages may cost no more than building pumped storage sites in the mountains, but no country has yet invested significant amounts of money in developing this approach.

Other ways of holding a reserve of electricity include making hydrogen—by splitting water into its constituent components—or compressing air. Using electrolysis to separate the hydrogen and oxygen atoms in water in times of abundant electricity, storing the hydrogen, and then burning it to drive a generator when electricity is in short supply is a plausible alternative to pumped storage. Similarly, we could use surplus power to compress large quantities of air, store it in depleted oil or gas reservoirs or salt mines, and then use the power of the air's expansion to drive generators when the power is needed. Both of these techniques look more expensive than water storage, though little work has yet been done to firmly quantify the costs. Pumped storage is also probably more efficient than these alternatives in that a greater fraction of the stored energy is recreated as electricity when it is needed. The downward flow of the water can recapture about 70 percent of the energy needed to pump the water upward in the first place. Compressed air and hydrogen manufacture don't look as though they can generate quite the same return, although the differences are not large. In August 2009, California's Pacific Gas and Electric Company announced a plan to build a trial compressed-air storage facility that could provide as much electricity as a medium-sized power plant for about ten hours.

Managing demand to meet supply
Supply and demand need to balance almost exactly on an electricity grid. Otherwise, the voltage or frequency of the alternating current would move outside the tolerances of home appliances and business equipment, possibly causing damage. Pumped storage is a way of quickly adjusting supply, but there are also methods to almost instantaneously reduce demand. In the jargon of the electricity industry, this activity is known as "load shedding." Some manufacturing companies, for example, have agreements that allow their electricity supplier to disconnect them at a few moments' notice. In return, they pay lower prices for

their electricity. This system works well. Although it is designed primarily to shave a little off the sharp daily peaks of electricity demand, there is no reason why the same approach couldn't be adapted to deal with temporary shortfalls in wind power at all times of day.

In the U.S., some electricity companies operate a slightly different scheme. They pay customers a rebate every month for promising to immediately reduce their electricity consumption when asked. For big commercial customers, this might work out to a savings of $100 per year for every kilowatt they commit to reducing. So if a company usually uses 400 kilowatts to power its office block but agrees to reduce consumption to 100 kilowatts at a few seconds' notice, it would receive $30,000 a year. These programs are voluntary, but they work well because most users can cut their demand easily with only minor inconvenience. Shops can reduce lighting use, hospitals can turn on their emergency generators, and businesses can shut off their air conditioning for an hour or so. To the electricity provider, these reductions may be equivalent to having an extra power station available; however, a spare power station that sits idle 90 percent of the time would be a much more expensive and polluting solution.

The major load-shedding programs around the world typically cover about 2 percent of peak power use, enough to cope with temporary energy deficiencies as long as wind is not too great a percentage of total electricity supply. For Spain and Denmark, countries where wind occasionally provides a very large fraction of total electricity demand for several hours at a time, this scheme would not be enough on its own. However, many people in the electricity industry think that load-shedding programs could cover 10 percent of peak power use or even more. This would be enough to protect the power grid from wind's small minute-to-minute variability in almost all circumstances.

Individual households can also be encouraged to reduce demand on a signal from the national grid. More and more countries intend to provide homes with "smart" electricity meters that can be remotely instructed to switch appliances off or that can limit total household power use to a set level—say, 3 kilowatts. Already, some French homes are fitted with meters that restrict energy consumption to this level. In Italy, almost all the customers of the main electricity company have smart meters and can reduce their bills by switching their electricity use to the times of day when prices are lowest. More advanced meters could be used to switch off non-critical machines such as dishwashers and washing machines at moments when wind power drops. The technology is already available to do this. A signal carried over the mobile phone network might trigger an electronic on/off switch at the wall socket of those electric appliances that use large amounts of electricity.

A program involving all three of these techniques for balancing supply and demand—better electricity grids, greater energy storage, and load shedding—allows electricity operators to deal with the occasional unexpected drop in wind power. There will be some costs, and some inconvenience, such as washing machines unexpectedly switching off for an hour or so, but it can be done.

In many parts of the world, regional electricity grids are already finding ways to reduce peak demand. In southern parts of the U.S., electricity suppliers often struggle to cope with summer afternoon peaks resulting from increased use of air conditioning. The arrival of large amounts of wind energy means that the problem of matching supply and demand is becoming more urgent: the grid has slightly more variability of supply in addition to fluctuating demand that can rise very quickly as afternoon temperatures increase. In the most locations, electricity companies are particularly interested in finding ways to increase the amount of load shedding available to them.

Other techniques for managing supply and demand will become available. As Chapter 6 explains, the batteries in electric cars could also form a vital buffer to keep the electricity system stable. When enough people own electric cars, their batteries will offer an extremely attractive alternative to other ways of matching short-term fluctuations in supply or demand. Four million car batteries, each providing 3 kilowatts of power, would match the maximum output ever achieved by all Spanish wind farms. Because the batteries of electric cars contain so much energy, the grid could use their power for several hours without affecting the vehicle fleet's state of charge by very much. Intelligent electronics, connected to the grid, could detect when power was needed and instruct parked, plugged-in cars to start supplying power. Conversely, if the grid was oversupplied with wind, the batteries could soak up the excess. When we have enough of them on the road, electric cars will have enough capacity to keep the whole electricity system stable. Proponents of renewable energy sources such as wind and solar should therefore also be committed enthusiasts of electric cars.

These measures can deal with the relatively small problem of very short-term and unexpected fluctuations in wind output. The absence of productive levels of wind for long periods is a more difficult problem, even when this absence has been predicted many weeks in advance and a country has access to large amounts of power from other regions. The possibility of days and weeks of low wind speeds represents a real challenge to the operator of the electricity grid. One response is to have a very large amount of unused generating capacity, probably using gas as a fuel source. These plants won't need to be ready without warning but need to be ready to start generating with perhaps a day's or a week's notice. Some grids are already investing in relatively inexpensive gas-powered turbines to provide this backup power, but there are other routes forward.

In most of Europe, the wind availability tends to be greatest when electricity demand is highest—in the early evening of the winter months—partly because there's simply more wind in

49

winter, when European electricity demand is at its peak. However, it's also because windy conditions in themselves tend to add to electricity demand. High winds increase the heat loss from houses. Those houses heated by electric appliances will usually need more electricity when wind energy is most available. In these parts of the world, the supply of wind-generated electricity is strongly correlated with the demand for power.

In some hotter countries, this happy relationship of supply and demand does not occur, because summer air conditioning is often needed exactly when the wind is not blowing. In these countries, the obvious climate-friendly solution is to install large amounts of complementary solar power to meet summer peak demand.

In fact, solar energy will likely be the best way of balancing wind supply in most countries because of the inverse relationship between the amount of wind energy available and the strength of the sun. For the countries of Europe, a logical mixture of power generation would see concentrated solar power from North Africa (see Chapter 2) providing much of the summer electricity, with wind and wave power—both most productive in winter— taking most of the strain in colder months. Tidal energy from geographically dispersed sites, generating maximum power at different times of the day, could provide a solid base of power availability, particularly at the equinoxes. In reserve, and ready to fire up if adequate supply looked uncertain, could be large numbers of wood-burning power stations. Unlike other forms of renewable generation, wood power plants need to pay for their fuel, but the cost of the generating equipment is not particularly high. It therefore makes sense to use wood as the main backup when all other renewable sources cannot match demand over periods of days and weeks.

We will still need gas and coal stations. So even countries with extraordinary resources of renewable energy, such as Scotland, with its wind and waves, or Canada, with its tides, will still need to work on capturing and storing the carbon dioxide emitted from the remaining fossil fuel power generation, as discussed in Chapter 8.

MORE WIND MYTHS

The opponents of wind energy focus not only on the perceived ugliness of turbines and the unreliability of the power, they also direct criticism at the potential impact on wildlife. Many of these concerns are unwarranted, and others can be exaggerated. Most land animals get used to turbines very quickly. Horses and cows, for example, ignore the rotating blades very soon after they are installed. Local birds are also largely unaffected by wind farms, although one wind farm in Norway has probably been responsible for almost wiping out a colony of rare eagles. Migrating birds may experience more of a problem, although the effect is still utterly insignificant compared with, for example, the impact of recreational hunting in Italy and France, road traffic, or predatory cats. More troublesome may be the impact on bats, which appear to be poor at avoiding the moving turbine blades.

Another frequently repeated criticism of wind power is the suggestion that the energy embedded in the manufacture and installation of a turbine is so great that it counterbalances the greenhouse gas reductions from several years of operation. This is simply not true. Research invariably suggests that wind turbines pay back the energy invested in them within a few months. Of course, the rate at which this happens depends on the windiness of the site and on the amount of concrete used to make foundations and access roads, but commercial wind power seems to have a strongly advantageous payback of energy and carbon dioxide. The one possible exception is that wind farms constructed in areas of thick peaty soils may result in substantial emissions of methane (a global warming gas) and CO_2 from the peat drying out and rotting. Good construction techniques that avoid undue disturbance of the surrounding landscape can prevent this severe problem.

One way of verifying this favorable general view is to estimate the carbon dioxide saved by using wind turbines and compare this figure with the emissions produced in making the steel for the turbine. We can do this very roughly by looking at the figures for Adam Twine's wind farm in southern England.

Each one of Twine's turbines will generate about 2.5 gigawatt-hours per year—enough to provide the electricity for perhaps six hundred local homes. If the wind farm hadn't been constructed, this electricity might instead have been generated at the Didcot coal-fired power station, ten miles away. Didcot would have emitted at least 2,000 tons of carbon dioxide by burning the coal necessary to produce this amount of electricity. One medium-sized wind turbine therefore saves these emissions every year.

How does this figure compare with the energy needed to make the turbine and its tower? Over 90 percent of the weight of a typical turbine is steel, and a large fraction of the total energy used to manufacture this steel arises from smelting the metal, usually in a blast furnace. An efficient modern steelworks emits about 2 tons of carbon dioxide per ton of finished steel. The weight of each of the turbines on Adam Twine's farm, including the blades and the supporting steel pole, is less than 200 tons. So the emissions from making the steel will be no more than about 400 tons of carbon dioxide. The cement in the concrete foundation and the groundworks to allow access to the wind farm will have added to this figure, perhaps bringing it up to 500 or 600 tons. Compare this number with the 2,000 tons of emissions that each turbine will save *every year*. By this calculation, the greenhouse gases arising from producing and installing the turbines will have been outweighed by the savings in emissions at Didcot within just four months of use. This informal calculation is not precise, of course, but it demonstrates that a wind turbine, in its lifetime, is likely to produce perhaps a hundred times the energy used in its manufacture.

One final concern critics occasionally raise about wind power is that erecting thousands of turbines might radically change local or global weather patterns by slowing down the speed of the air. This worry might be valid if turbines captured more than an infinitesimal share of the total energy in the wind moving around the world. Any significant change in global weather patterns would probably only occur if a measurable fraction of the world's surface were devoted to wind farms. Today, the reduction of wind speeds as a result of new turbine construction is almost certainly less than the increase in wind levels caused by the world's loss of forested area. Trees slow down the wind, too.

SOLAR ENERGY

Enough to power the world many times over

THE SUNLIGHT hitting the earth's surface every day contains around seven thousand times the energy in the fossil fuels that humanity consumes. If we could find an economical way of exploiting this energy, then all the world's energy and emissions problems would be solved. Even with today's technologies, solar collectors on less than 1 percent of the world's unused land could comfortably match all fossil fuels in the energy they provide. Which is no surprise when you consider that a sun-soaked tropical area of just 100 square feet—approximately the floor space of a small bedroom—receives as much energy from the sun as the typical global citizen consumes for transport, heating, food, electricity, and all other aspects of life.

Of course it isn't as simple as that. For one thing, the sun's rays are at their most powerful in the tropics, while much of the world's population is in temperate countries thousands of miles away. But the potential is huge, and solar technologies have many advantages. Not only are they climate friendly, but they're also non-polluting and almost noiseless, and they require little maintenance. In addition, unlike biomass energy, they make use of non-productive space—be it deserts or urban rooftops—and therefore don't put pressure on food production.

There are three main ways to capture the sun's energy. The first is to put long tubes containing liquids in direct sunlight. The liquid in the tubes gets hot and, with a heat exchanger, can be used to heat water for showers or for washing clothes. The second way is to use panels of photovoltaic (PV) cells to turn the photons of light directly into electricity. Finally, there are solar concentrators, which use mirrors to focus large amounts of sunlight onto a small area, intensively heating fluids and using their energy to drive a turbine or a Stirling engine to generate electricity.

The first of these approaches—solar water heating—has been available for centuries. It is a straightforward technology and can be remarkably efficient. On our house in cloudy Oxford, we have forty glass tubes about 6 feet long on the roof. Inside each tube is a thin, flat foil of copper. This foil is heated by the sun and transfers the heat to a liquid in a thin pipe running in the center of the tube. This liquid is pumped into a heat exchanger that transfers the energy to the hot-water tank. It is an extremely simple and reliable system. On a summer's day in southern England, it provides all the hot water for five people. It captures perhaps 70 percent of the light and infrared energy falling on the glass tubes and transmutes it to useful hot water. No other solar technologies are anywhere near as efficient. But heating water for baths and showers is not a large part of the energy needs of most households. In Europe, heating hot water demands less than 1,000 kilowatt-hours per person each year, or less than 3 percent of total energy requirements. Nevertheless, in sunny countries, solar hot water makes good financial and environmental sense. Cheap solar collectors can provide heat for hot water for most of the year. In high-latitude countries such as Britain, I have to admit that solar water heating will barely cover its cost, even at today's fossil fuel prices.

To get the greatest impact from solar energy, we should use it to generate electricity, not hot water. Indeed, we need to get solar electricity to the point where it offers developing economies a cheaper way of fulfilling their growing needs for power than by burning fossil fuels. Higher-latitude countries may find wind energy the cheapest form of power, but solar power will eventually be the best way of generating electricity in the tropics, where wind speeds are generally lower.

The traditional way to produce electricity from sunlight, by means of photovoltaic cells, is well established but still expensive. This chapter explores the various ways that scientists and engineers around the world are trying to bring down costs and raise efficiency. Some of their innovations are very promising. However, the chapter also argues that mirror-based solar concentrators are just as significant and offer huge potential. Prince Hassan bin Talal of Jordan, a leading backer of this technology, outlines the vision:

> In deserts, clean power can be produced by solar thermal power plants in a truly sustainable way and at any volume of conceivable demand . . . This gives the deserts a new role: Together with the many other forms of accessible renewable energy the newly utilized desert would enable us to replace fossil fuels and thus end the ongoing destruction of our natural living conditions.

The idea of using the sun in the Middle East and Africa to provide Europe with limitless and cost-competitive power is hugely appealing. Networks of enthusiasts for this project have sprung up all around the region, led by scientists and electric power utilities in Germany. Solar power concentrators have immense potential around the world. The arid southwestern U.S. is particularly suitable, but any area of sunny desert will provide huge quantities of energy, potentially at relatively low cost. As we'll see later in this chapter, the challenges are primarily logistical and commercial, not technological.

SOLAR PHOTOVOLTAICS

In 1958, the U.S. launched Vanguard, the first satellite equipped with solar photovoltaic cells to provide electric power. The panels produced about 1 watt of electricity, not enough to cover the passive use of electricity in an idle TV set today. Only a minuscule fraction of the energy hitting Vanguard's solar panels was turned into electric current. Half a century later, advances have taken the efficiency of some commercial photovoltaic panels to a maximum of just over 20 percent, though the cost remains oppressively high.

Simply put, a photovoltaic panel creates electricity when light energy (a photon) hits the silicon surface and pushes an electron out of the top layer of the silicon and across an electrical junction inside the panel. The movement of this electron creates a useful voltage. When wires are connected to produce a circuit, this voltage means that current will flow, eventually taking the displaced electron back to the top layer. Solar cells work best in strong sunlight but will also generate some power on an overcast day from the diffused light that gets through the clouds. They're not particularly complex devices, but the technical challenges in producing them cheaply are formidable.

Most solar panels manufactured today are made from expensive slabs of extremely pure silicon. The silicon is derived from a very abundant substance, common sand, but the process of refining it and ensuring that it is pure enough for electrical use is complex and energy intensive. As the solar panel industry grew, encouraged by enormous subsidies in Germany and other countries, the supply of pure silicon did not keep pace. Significant shortages in 2008 pushed the price of solar panels up. But the price rise was followed by a sharp decline as large numbers of new factories in China and elsewhere began producing unprecedented volumes of silicon in 2009, and some of the major markets, such as Spain, saw sharp reductions in the financial incentives to install PV systems. By the last quarter of 2009, the prices of PV modules were more than 20 percent below the levels of mid-2008.

Solar panels can be built in any size. Small rectangles of silicon provide the power for personal calculators and other minor domestic appliances. Much bigger blocks are used to make conventionally sized solar panels, which are over 3 feet tall and somewhat less than 3 feet wide. In a solar power station, huge numbers of these panels can be chained together, all providing electricity and perhaps generating as much as 50 megawatts in full sun. As solar panels decline in price, which should happen quickly over the next ten years, we can expect to see them installed in larger and larger groups, with total power output close to that of conventional power stations.

By early 2009, the installed photovoltaic panels across the globe could produce about 15 gigawatts of electricity if working in full sun, somewhat less than 10 percent of the worldwide capacity of wind power. (The power of a solar panel is usually expressed as the maximum output when the sun is shining strongly at midday.) Most of the time, the electricity actually produced will be much less. In a very sunny country, a day's electricity from a panel with a rated power of 1 kilowatt might be about 6 kilowatt-hours. This is about a quarter of the electricity that would be generated if the panel were working at peak efficiency for the full twenty-four-hour period. This typical performance implies that all the solar photovoltaic panels in the world currently provide less than a tenth of the total electricity demand of a country the size of France or little more than 10 percent of the needs of the state of California.

Solar electricity is growing rapidly, perhaps by 30 or 40 percent a year, but today's global photovoltaic output is only equivalent to a couple of very large coal-fired power stations or a small cluster of nuclear plants. This comparison helps show the enormous scale of the challenge. After fifty years of research and development into photovoltaic technology, we are still only obtaining a small fraction of 1 percent of world electricity demand from solar sources. So why should we be optimistic that all this will change and that solar energy will provide a significant percentage of world electricity within a decade or so?

THE COST OF PHOTOVOLTAICS

The financial performance of solar panels has consistently improved over the years as technical progress has reduced costs and raised the output of electricity. But PV is still a very expensive way of generating electricity except in the sunniest places. If installed today, the large and slightly ungainly slabs of silicon on the roof of my home and the associated electronics attached to the wall of the garage would cost about $15,000. In roughly four years on a house in Oxford, they have produced about 6,000 kilowatt-hours of electricity. If I'd sold that power in the wholesale market for electricity, I would have banked about $500. The panels will probably last another twenty-five years before their performance begins to degrade, so the total value of their power output without subsidy will almost certainly never cover the original cost. In our case, a government-sponsored capital grant available at the time of installation (but not now) and the enhanced prices we obtain for the electricity we export mean that the panels will actually earn our household a reasonable return of at least 5 percent a year on the capital we invested.

On the roof of a house in high latitudes, photovoltaics don't seem like a cheap way to reduce carbon emissions. Partly, this is because putting panels on a tall house requires expensive scaffolding and several days' work. It's also because the two "inverters" needed to change the low-voltage direct current output from the panels into the alternating current required by the electricity system contain sophisticated electronics. Each inverter cost over $1,500 four years ago, and the total amount of power we have generated thus far wouldn't even cover the cost of these devices.

Germany introduced generous "feed-in tariffs" in 2000, aimed at encouraging faster installation of renewable energy. Property owners who put solar panels on their roofs are entitled to substantial payments for every kilowatt-hour of electricity that they feed into the local electricity grid. The incentives were, and still are, enormous. Each kilowatt-hour of solar energy is currently worth about 45 euro cents, or six times the typical wholesale price for electricity. The feed-in tariff set the solar photovoltaic industry alight in Germany, and, by the end of 2007, over 300,000 homes and businesses had solar roofs. California, by comparison, had only about a tenth of this number in early 2009. The cost in Germany has been enormous—probably more than $7 billion a year for less than 1 percent of its total electricity need—but the subsidy has built expertise and knowledge. It is little exaggeration to say that the world solar power industry would barely exist without the subsidy to German homeowners.

The International Energy Agency criticized the expense in a 2007 review. It said:

> Estimates show that between 2000 and 2012, the feed-in tariff will cost 68 billion euros [over $100 billion] in total. In particular, the subsidies provided to solar photovoltaics are very high in relation to output; they will eat up 20 percent of the [renewables] budget but contribute less than 5 percent of the resulting generation.

High payments to the owners of solar panels might have cost other German homeowners large amounts of money, but there is little doubt of the effectiveness of the policy in helping photovoltaic manufacturers across the globe reduce their costs and improve their production processes. At one stage, over half the solar electricity produced in the world came from German roofs, even though the average German solar panel produces substantially less than half as much electricity as the same panel would in the Sahara. Those of us not living in Germany should be grateful for the generous solar subsidy and its impact in bringing manufacturing costs down across the world.

The billions of dollars spent overpaying Bavarian farmers to put solar panels on their cowsheds have attracted widespread political attention, and other countries have copied the idea of feed-in tariffs. France and South Korea, to give just two examples, are following the German lead. But no country wants to subsidize renewable energies forever, and panel manufacturers around the world believe that solar photovoltaic technology must rapidly become competitive with fossil fuels if it is to continue to prosper.

As a result, all the competing companies vying to build big businesses in solar technologies have one target in mind. For solar power to become truly competitive with fossil fuels, these businesses say that photovoltaic panels have to cost the customer no more than about $1 per watt of maximum power, or about $1,000 per kilowatt of peak power. The costs of the associated electronics and the expense of the installation will probably double this price, taking it up to $2,000 a kilowatt. This figure is about a fifth of the cost of the equipment installed on the roof of our house four years ago, so the challenge is enormous.

Of course, solar doesn't generate much electricity when the sun isn't shining, so the dollar-per-watt figure isn't easy to compare with the cost of a coal plant or natural gas power station. In a sunny region, solar panels with a capacity of 1 kilowatt will generate over 2,000 kilowatt-hours a year. Wholesale electricity prices vary around the world, but this amount of energy typically might be worth $200 to $250 to the power producer, and much less in some markets, including much of the U.S. Since solar energy has very low yearly operating costs, the payback period on the initial investment of $2,000 per kilowatt might be as little as eight or ten years in countries with expensive electric power. If solar photovoltaic technology is any more expensive than this, it will require continuing subsidy. The big U.S., Chinese, and Japanese panel manufacturers are only too aware that the overgenerous "feed-in" prices paid for solar energy exported to the national electricity grids in places such as Germany are already edging downward. Electorates will not be willing to support high-price solar electricity indefinitely. Very sensibly, these manufacturers are therefore aiming to get their production costs down to a level such that customers will not need subsidies to justify buying photovoltaic panels.

The "dollar a watt" target is also known as "grid parity"—that is, the point at which the industry believes that using solar power to supply the electricity distribution grid is no more costly than using fossil fuel power plants. Progress toward this objective is surprisingly fast. One recent industry study said that solar technologies will be competitive with coal and gas by 2015 across most of the U.S., excluding only the least sunny areas.

The largest U.S. solar panel manufacturer, First Solar, is even more optimistic. It expects to achieve grid parity no later than 2012, and several other companies, some using very different technologies to make their panels, have made similar claims. By 2012, First Solar will probably be supplying several gigawatts of PV panels each year to its customers, largely to companies building huge farms of solar panels in sunny areas. Photovoltaic technologies tend to work less well in high temperatures, so these solar power stations will have to be located in areas that receive good solar radiation but that are relatively cool. South-facing mountain slopes are ideal.

Having to drive down costs 20 or 30 percent a year is one problem for the relatively small number of large panel manufacturers. The other issue they face is even more challenging. At least four different types of solar panel exist, each using different semiconductors in a variety of thicknesses. The old

63

technology—heavy layers of pure silicon—is under threat from upstart new approaches, often using very thin coatings of semi-conductors on a simple backing material. In this exciting but still obscure industry, one of the world's great business battles is just beginning. Not entirely friendly controversy rages between man-ufacturers over which method will ultimately prove to be the best way to compete with fossil fuels. Most of the companies building their advanced new factories have committed irrevocably to one technology or another. Many billions of dollars ride on success.

Which of these options will give us electricity at the lowest price per kilowatt-hour? And which will give us the fastest rate of increase in the total generating capacity installed around the globe? "Old-fashioned" panels, manufactured from large amounts of silicon, have the advantage of capturing a relatively large percentage of the energy of the sun, but they are expensive to manufacture. Is it better to focus on much thinner films of sili-con that deliver smaller amounts of electricity from every panel but at a substantially lower initial cost?

The next question is whether silicon is the best material at all. First Solar has opted for a semiconductor called cadmium telluride, from which it makes thin panels that are relatively inexpensive but only moderately efficient at capturing the sun's energy. First Solar's cells convert about 10 percent of the sun's energy into electricity, though the company is planning improve-ments that will take this figure up to 12 percent or more within a few years. Other companies, including the secretive Nanosolar, funded in part by the billionaire founders of Google, are con-centrating on another semiconductor material, known as CIGS (copper indium gallium diselenide). Nanosolar's backers are hoping that its revolutionary technology, which simply "prints" the semiconductor material onto a flexible metallic backing layer, will prove to offer panels of such low cost that they will be wrapped around the exterior of millions of buildings across the world. As its name indicates, Nanosolar is using nanotechnol-ogy to precisely arrange the atoms on the printed semiconductor surface. Its frequent claims that this approach will eventually produce extremely cheap panels are convincing to many outsiders but treated with undisguised skepticism by other businesses in the PV industry. After several years of R&D, the reclusive company finally shipped its first commercial panels to a solar farm in Germany in the last days of 2007. In September 2009, it opened its first factory in Germany able to continuously produce solar panels at a rate of one every few seconds. If Nanosolar's most important boast—that it can print solar PV cells of one hundredth the thickness of conventional silicon panels at speeds one hundred times as fast as current manufacturing processes—is even partly true, the cost of PV is likely to fall extremely quickly as it ramps up its production.

The battle is not yet over. Some large manufacturers are sticking with conventional thick silicon solar panels, believing that the price of pure silicon will eventually fall dramatically, meaning that their raw materials will be much less costly in the future and it won't matter if they use several pounds in each panel. The Japanese company Sharp, which has been making solar panels since 1959 and has a large market share of solar installations around the world, is focusing its efforts on improving the conversion efficiency of its silicon panels rather than reducing the amount of silicon it uses.

The competition between all of these alternatives is essentially between those who believe that reducing the cost per square foot of panel is the most important objective and those who think it's more important to focus on improving the efficiency of capturing solar energy. Companies focusing on thin films, whether of silicon, cadmium telluride, or CIGS, are betting that the vital change is getting panel prices down to low levels. If the panel is very cheap to make, then it doesn't matter much if you have to use a larger area.

On the other side of the argument, those concentrating on improving the efficiency of light-to-electric conversion think that panel costs are not the most important element. They claim that

65

the cost of panels, both thin film and conventional, will eventually fall substantially but point out that this cost is only a part of the bill for a solar installation. Eventually, they predict, the panels will represent less than half the total cost of putting a solar roof on a large commercial building, with labor, cabling, and inverters accounting for most of the rest. Therefore, the argument goes, the most important aim is to get the largest possible efficiency per area of roof and per dollar of installation costs.

The companies betting that the future of PV is based around common silicon attack their competitors on another front. They say that cadmium telluride and CIGS companies both need reliable supplies of chemical elements that are in dangerously short supply. First Solar, the hugely successful business based on cadmium telluride, frequently has to rebut assertions that the world is simply going to run out of tellurium in the next few years. Although only a few grams of tellurium are needed for each panel, the cost of this rare metal has risen dramatically in the last few years. First Solar says it has guaranteed supplies for several years ahead. Other supporters say that tellurium is available in very large quantities on underwater ridges beneath the oceans where it can be cheaply mined.

CIGS is the subject of a similar debate. Skeptics make pessimistic statements about future shortages of the indium metal in CIGS. Here the issue is not so much absolute scarcity: indium is at least as common in the earth's crust as silver is. But the current generation of LCD screens, used for computers and televisions, is competing for the relatively small amounts of the metal currently mined every year, and the price has risen sharply. Nevertheless, indium prices probably won't be a substantial problem for CIGS manufacturers in the future. Within the next ten years, video and computer displays will likely be made using a technology that does not use indium. The CIGS manufacturers are gambling that indium will become cheaper as supplies increase and demand from competing products eventually falls away.

There is one further complication. The maximum level of electrical efficiency for standard panels is about 20 percent. (Thin-film panels struggle to reach 10 to 12 percent.) This low figure arises because current PV panels are only able to capture the energy from a small portion of the visible light spectrum. Red light passes through the panel without dislodging electrons, whereas blue light is largely reflected. In theory, a PV device that combined several layers, each with different absorption characteristics, could capture far more energy and perhaps even exploit invisible infrared light. Panels like this are called "multi-junction" cells, and they've already been demonstrated in laboratories with efficiencies of nearly 40 percent.

The ultimate aim must be for the manufacturing companies to make cheap, thin, printed, multi-junction solar cells, probably from inexpensive silicon. When this happens, we will see buildings around the world covered in these panels and generating all the electrical energy that they need at a price to beat any electricity delivered from large, remote power plants. The crucial question is when. The amount of private capital going into PV technology is large enough to achieve the objective of very low panel prices, but my suspicion is that mass availability is as much as ten years away. Although all the major photovoltaic manufacturers are publicly saying that advances in cost and performance are going to be extremely rapid, I think a little skepticism is probably justified.

Given the slow pace of progress in PV over the past half century, why should anyone be even this optimistic? First, nanotechnology really does make a difference. Now that companies can make specialized materials whose atoms are very precisely arranged, we are seeing rapid advances in the ability to capture the energy of the photons hitting the panel. Second, the impact of the German feed-in tariff has been to vastly increase the total number of panels being made around the world. The effect on manufacturing costs, ignoring the temporarily very high price of silicon, has been dramatic. The world is currently only making a few gigawatts of PV panels each year, but we are doubling the accumulated manufacturing volumes every couple of years. The cost reductions achieved so far from moving down the learning curve give us good reason to believe that as volumes continue to increase, we will see continued very sharp declines in cost.

Eventually, PV will almost certainly be the technology of choice for small-scale and localized electricity generation in sunny countries. With luck, low-cost solar panels will be available to meet the needs of remote communities in Africa and Latin America well away from the electricity grid. In other words, these places may never need to install fossil fuel power stations. All they would need is a low-cost storage technology to cover nights and periods of cloudy weather. This bypassing of fossil fuels is already happening on a very small scale. The German company SunTechnics is supplying panels to Namibia, where many of the people live far from a reliable electricity supply. The electricity users do not buy the solar panels and other electronics but simply prepay for the electricity that they use. The utility company that operates this service is, in effect, renting the solar kit to the household or business and can move it elsewhere if the customer no longer wants the power or turns out to be a bad credit risk.

In developed countries, solar PV will eventually make most sense installed on the user's premises, rather than in the huge centralized power stations that First Solar is currently focusing on. The primary reason for this is that a commercial solar farm feeding into the national grid will get paid the wholesale prices for power, which are typically about 50 percent of the price paid by homeowners or small commercial customers. By contrast, solar PV installed on homes or offices displaces power that the building user would otherwise have purchased at the retail price. PV is one of the few electricity sources that can be installed on a very small scale and still be reasonably productive. A wind turbine on a house costs ten times as much as a commercial wind farm per unit of electricity generated, but the comparable ratio for solar PV is probably only about two. This means it may eventually make good financial sense for a building owner to put PV on the roof, displacing electricity supplied by a utility. The capital cost disadvantage of a small installation does not outweigh the savings from not having to pay retail prices for power.

The largest problem, as with some of the other technologies discussed in this book, is scaling up solar panel manufacturing quickly enough to dent global greenhouse emissions in the short window of time available. Perhaps twenty or thirty companies in the world currently produce large numbers of advanced PV panels or hope to be manufacturing them in the near future. First Solar hopes to be making enough panels in 2010 to generate 1 gigawatt in peak sun. This is less than the new capacity of coal-fired power stations being installed in China every four days. All the world's manufacturers of solar panels added together are likely to produce about 12 gigawatts of new panels a year by 2010, barely enough for six weeks of grid expansion in China.

So can PV ever become a technology that supplants a significant amount of fossil fuel generation? Demand for electricity is increasing by 3.5 percent a year outside the industrialized countries. The small number of companies with the technology and experience to make competitively priced PV are going to struggle to make enough panels to cover the worldwide growth in electricity demand over the next few years. But we shouldn't be too pessimistic: the scope for wholly unexpected and truly revolutionary advances in photovoltaic technology is at least as great as any of the other technologies discussed in this book. If Nanosolar or one of its competitors does find a way of printing huge volumes of cheap semiconductor materials that can be easily added to the exterior of most buildings, the scope for photovoltaic technologies to change the world is almost unlimited.

Once the technical problems have been solved, the way is clear for PV. The environmental consequences of photovoltaics

69

are limited, and objections to the appearance of panels on the roofs of buildings or in large farms are few. The cadmium telluride used in First Solar's and some other manufacturers' panels is toxic but presents few dangers when in use in solar installations. Claims that thick silicon panels embed more energy than they are ever likely to capture from the sun occasionally resurface but are not supported by the research carried out into the energy balance of PV. Studies some years ago suggested that panels repaid the energy cost of making them within about three and a half years, but advances in manufacturing efficiency and in the amount of light captured by PV mean that the energy payback period is now probably only about two years. Since the panels will normally last over twenty-five years, the return is good. Thin-film panels have even better energy balance because they require far less energy to make.

CONCENTRATED SOLAR POWER

Photovoltaic cells directly convert photons from sunlight into electricity. The other way of generating power from the sun is to concentrate the rays onto a liquid. The liquid heats up and can be used to boil water, which then forces its way through a steam turbine, generating electricity. The efficiency of this process, expressed as the percentage of the sun's energy converted into electricity, can be greater than with a PV panel. Steam turbines are the method of generating electricity used in all existing coal and nuclear plants, so we understand well how to convert heat into movement and then into electrical energy.

This form of electricity generation is now usually called "concentrating solar thermal power," often shortened to CSP. This name covers perhaps five or more separate approaches. The newest to arrive in a commercial application is known as "solar towers." At the first working example, near Seville in Spain, six hundred mirrors placed in a circle reflect concentrated sunlight onto a single point at the top of a specially constructed tower. At present, this tower generates only about 11 megawatts—equivalent to three or four large wind turbines working flat out—but as more mirrors are added, the power will increase.

Another solar thermal technology is the solar dish. Looking like a huge satellite receiver from the 1960s, this apparatus tracks the sun as it goes across the sky. The mirrors on the interior of the dish reflect sunlight toward a focal point. At this point there is a Stirling engine, a machine that turns the expansion and contraction of gases into power by turning a crankshaft. At the moment, there are few working examples, and investors are still to be convinced of the financial potential of this approach to solar energy.

The best-established solar thermal technology is slightly different. This form of CSP uses long parabolic troughs covered with reflective material to concentrate the sun's powers onto a thin tube, called a receiver, at the focus of the parabola. A good solar collector can focus about a hundred times the usual power of the sun onto the receiver. The receiver contains water or, more usually, oil. At a new large project near Granada, Spain, the thin tube of oil is heated to over 400°C (750°F) in full sun. The hot oil is passed through water, with which it exchanges heat. The water rapidly heats up, boils, and then turns into energetic steam, ready for powering a rotating turbine, in exactly the same way as it would in a coal-fired power station.

The Granada power plant, called Andasol 1, is one of the first of what its German proponents hope will be tens of thousands of similar installations across the sunniest parts of the world. It is sizeable, covering an area of about 125 acres. Several features help maximize its usefulness as a generator of electric power. The parabolic reflectors run in north–south lines, and, using small electric motors, the mirrors are rotated from east to west during the day, ensuring that they will face directly into the sun during the daytime. The plant also uses molten salts to store heat in order to extend the operating day to seventeen hours or more. During the sunniest part of the day, part of the heat that is collected is used to melt the simple salts (potassium and sodium nitrates). The retained heat is then used to create the steam needed to power the turbines when the sun is down.

The new Andasol plant generates about 50 megawatts when it is working. Over the course of the year, it will deliver about 180 gigawatt-hours of electricity, providing enough for about fifty thousand European homes. To put this figure in a slightly pessimistic perspective, we would need about thirty of these plants to provide as much electricity as we get from just one coal-fired power station. Although interest in this technology is growing rapidly, it is from a low base: one calculation suggested that only thirty large CSP projects were in active development around the world in early 2008. In September 2009, 800 megawatts of capacity was under construction in Spain, and about ten times as much was in the early stages of permitting.

Many countries are extremely receptive to the Andasol-style approach to CSP. A similar power station was completed in February 2008 in Nevada. It also uses troughs of parabolic mirrors arranged in long rows. The Spanish construction company that built Nevada Solar One proudly claims that it has 48 linear miles of parabolic collectors with 182,000 separate curved mirrors that focus the sun's energy onto eighteen thousand absorbing tubes. It covers an even bigger area than the Granada power station.

Very sensibly, Nevada Solar One was built very near a long-distance power transmission line, meaning that it was relatively simple to connect to the electricity grid. In Nevada, and much of the southern U.S., electricity demand peaks in summer afternoons, as air conditioning is running at full power. The high demand generally means that spot prices for power are also at their maximum at this time. Electricity traded in the wholesale market during this period can cost several times as much as the electricity traded in the cool of early morning. Usefully, a summer afternoon is when a solar power plant is also producing the most electricity, meaning that its output commands a premium price.

Utility companies in the parts of the world facing power shortages on hot afternoons are likely to be particularly keen on CSP plants. In February 2008, an Arizona electricity supplier announced a plan to build the world's largest single solar thermal installation, about 70 miles southwest of Phoenix, and a smaller Tucson plant was announced in early fall 2009. The 280-megawatt Phoenix plant will be operational in 2011 and will triple the amount of renewable electricity now produced by its owner. When fully operational, the plant will provide the electricity for seventy thousand power-hungry Arizona homes.

By comparison with most other renewable technologies, CSP plants will not be particularly energy intensive to construct. But even the relatively simple solar collectors use a lot of steel. The Arizona plant—called "Solana," Spanish for "sunny place"—will use 80,000 tons of the metal. Done in an efficient blast furnace, the manufacture of this steel will cause the emission of perhaps 160,000 tons of carbon dioxide. It will take at least six months of operation for the plant to pay back this carbon debt. Nevertheless, this ratio is better than for conventional PV panels and about the same as wind turbines.

Solar concentrators also have the enormous advantage of being relatively simple and reliable. The first CSP plants were built in California about twenty years ago and have worked well since then. The total output of these remarkable power stations in the Mojave Desert is six or seven times higher than that of Nevada Solar One or Andasol. They have a good record of reliability and are expected to last at least another fifteen years. Surprisingly, only recently have investors come to see the advantages of replicating this successful experiment elsewhere in the world.

Perhaps equally importantly, CSP doesn't require scarce metals, so its growth won't be held up in the way that First Solar's cadmium telluride PV cells may be. It doesn't require expensive silicon, so it will escape some of the problems of conventional polycrystalline panels. It is almost completely non-polluting, and, at least in theory, we can build multiple plants very quickly.

What is holding it up? Why are we not seeing hundreds of CSP plants in construction across all our hot deserts? Two reasons predominate. First, the current generation of parabolic dish reflectors is expensive. The total cost of the new Granada plant is about $8 per watt of peak capacity, while the Nevada station is about $5, both well above grid parity. The solar PV manufacturers are aiming for a figure of about $1 for every watt of midday generating capacity and about another dollar for the associated electronics. CSP plants deliver much more energy in the morning and afternoon (because they can follow the sun, whereas most PV is fixed on roofs), but this comparison shows that CSP still has some way to go to be competitive with fossil fuel technologies.

Nevertheless, I think we can be optimistic that CSP will decrease in price at a similar rate to the falling price of solar PV panels. The troughs of parabolic mirrors are the largest portion of the cost of a CSP installation. Very large-scale manufacture of these parabolas will reduce their cost dramatically. As the world accumulates more knowledge of how to manufacture the other components of solar thermal installations, we can expect rapid decreases in their price. The mirrors have to be very precisely engineered to focus the sun's energy accurately on a thin tube of oil, but this is a manufacturing rather than a technological problem. My guess is that costs will respond quickly as the CSP industry grows.

Some of the photovoltaic manufacturers are highly skeptical about the competing technology. They point to the need to install new electricity transmission cables to take electricity from remote deserts and also to the need for large amounts of water for cooling. Water is not generally easily available in the hottest areas. Despite challenges such as this, the U.S. Department of Energy gave an upbeat assessment of CSP in 2007:

Existing CSP plants produce power now for as low as 12 cents per kilowatt-hour (including both capital and operating costs), with costs dropping to as low as 5 cents per kilowatt-hour within ten years as technology refinements and economies of scale are implemented. Independent assessments by the World Bank, A.D. Little, the Electric Power Research Institute, and others have confirmed these cost projections. While not currently the lowest cost electricity, CSP is already close to competitive in peaking markets.

By "peaking markets," the Department means the time when electricity is most expensive and most in demand—that is, late afternoons in summer in most of the southern United States.

A small group of determined scientists and engineers has been working to excite policy-makers with a grand plan for CSP. Trans-Mediterranean Renewable Energy Cooperation (usually known as TREC) is pushing for almost unlimited adoption of solar thermal power. Its backers include governments from North Africa and the Middle East. As quoted in the first few pages of this chapter, Prince Hassan bin Talal of Jordan supports this technology for his country, which consists largely of hot desert. TREC's view—and nobody has ever stepped forward to contradict it—is that putting huge developments of solar troughs in the sunniest parts of North Africa and the Sahara could provide all the electricity that Europe and the Middle East need. The cost per kilowatt-hour, including a profit margin, would be competitive with electricity made from coal. TREC mentions figures eventually as low as 4 to 5 euro cents per kilowatt-hour, very similar to the U.S. government's estimates of future costs. At these levels, CSP plants are likely to be financially very successful. If we had started pushing CSP twenty years ago, it would now probably be producing electricity at a lower price than any other technology. This should be a lesson for us.

TREC claims that the whole of Europe, North Africa, and the Middle East could get its electricity from CSP plants on as little as 0.3 percent of the desert land area of the region. In one of the more telling illustrations of the power of the idea, TREC's website has a map of the Sahara on which a tiny red square is superimposed. All the electricity for the entire European region could come from power plants that would fit into this area.

But here we come up against the second problem that CSP faces. Unlike solar photovoltaics, whose effectiveness degrades at high temperatures, CSP will produce more electricity in the hottest weather. But getting the power out of a hot and uninhabited North African desert is likely to be difficult. A successful CSP plant needs to be close to a high-voltage electricity line. The very sunniest areas of the world, such as the Sahara or some of the American deserts of the southwest, have no high-voltage power cables at all. The Sahara is a good place to generate the electricity, but we need a way of getting the electrons all the way to where they are needed—the populated areas of Europe, across the Mediterranean sea.

Moving electricity long distances is troublesome. Building the power lines is expensive, but a more significant problem is that a substantial amount of the electricity is lost in transmission. With conventional alternating current (AC) transmission lines, too much of the power would disappear as dissipated heat or as waste electromagnetic radiation on the way to the place using the electricity.

CSP enthusiasts have an immediate response to this difficulty. They say that we should be using high-voltage systems with direct current (HVDC) transmission. The losses in HVDC systems are much smaller than in conventional high-voltage AC systems. A transmission line from North Africa to northern Europe, a distance of several thousand miles, would lose less than 10 percent of the electric power.

But are such transmission lines feasible? The electricity would have to go from, say, the Tunisian desert across the Mediterranean to Sicily and then northward through Italy. The distances are long, and the terrain will sometimes be extremely inhospitable. Getting to the Tunisian coastline should present few problems. Underwater cables would then be needed to cross the sea. The longest undersea power line today is the just-completed 360-mile link between Norway and the Netherlands. A huge cable weighing 70 pounds per foot carries HVDC between the two countries, enabling their electricity grids to exchange power. The cost of this cable, finished in April 2008 and fully operational a month later, was about $900 million, or about $2.4 million a mile. The line allows 700 megawatts to flow either way, but the expectation is that most of the power will come from Norway's hydroelectric stations into the European power grid. A cable from Tunisia to southern Italy would cover a much shorter length, and the sea conditions would make installation much easier.

So the distances are feasible, though the cost of the cable running along the floor of the Mediterranean will be high. Nevertheless, the world's electricity industry is used to this scale of investment. In fact, direct current transmission lines above ground may be cheaper and simpler than the alternating current transmission lines running across Europe's landscapes today. HVDC transmission towers can actually be smaller and less visually obtrusive than their conventional AC equivalents.

New onshore HVDC links are getting longer all the time. India and China's need for electric power is surging, and both countries are installing several major long-distance lines. In February 2008, the German power company Siemens completed a 750-mile HVDC link from power plants in western China to the industrial province of Guangdong. The transmission lines carry about 3,000 megawatts—more than the output of one of the very largest coal-fired power stations. Siemens is also building a long HVDC line carrying 5,000 megawatts, and its competitor, the Swedish/Swiss company ABB, aims to complete a 1,250-mile, 6,400-megawatt link, also in China, in 2011.

How many lines of this type would be needed to satisfy Europe's electricity demand with electricity from the Sahara? Germany's peak demand is about 100 gigawatts. Let's assume CSP from the Sahara eventually provides 50 percent of the country's electricity demand, with the rest generated by local wind, tidal, and other renewable supplies. Germany would therefore need about eight of these long-distance HVDC links from the desert, each with a typical length of perhaps 1,800 miles.

To provide the whole of Europe's electricity would probably mean at least thirty different transmission lines of the same size as the biggest direct current links being built today. One estimate has them costing over $3 billion each, including the portion under the Mediterranean. But, in itself, this problem is not insurmountable. A new nuclear plant may cost $8 billion or even more. Nevertheless, analyzing how the HVDC links could be built and financed is a challenge that has so far received too little attention.

The TREC concept is immensely attractive, and not just for Europe. Desert-based CSP could provide cheap, carbon-free electricity for the countries of North Africa and the Middle East. These states could use the power for industrial development and, perhaps most importantly, to desalinate water, thereby allowing a major expansion in the area of irrigated crops as well as improved availability of drinking water. The proponents of CSP also believe that the areas underneath the parabolic troughs will be very productive places for horticulture. The shading effect will improve yields and reduce the otherwise excessively high temperatures.

Algeria is one of the first North African countries to back a CSP project. The Spanish construction company Abengoa, perhaps the most enthusiastic proponent of concentrated solar power, is building a power station there using parabolic troughs. This plant will be able to produce electricity for the entire day because the turbines can be switched to burn natural gas at night. This advance is useful because it means that the solar

plant can operate as what is called "baseload"—reliably producing a steady stream of power at all times and in all weather
conditions. Most renewable technologies do not offer this security to the electricity companies. Some, such as wind power, are
unreliable, or their power is cyclical. Tidal barrages, for example,
generate most electricity just after high tides and none at all at
low tide. So the Algerian plant is providing a useful prototype of
how we can make CSP an integral part of the power grid. Algeria alone is talking about installing 6 gigawatts of CSP capacity,
equivalent to three very large coal-fired power stations.

Most assessments of CSP agree that the costs per kilowatt-
hour are likely to decline to below the figures for fossil fuel plants.
A substantial carbon tax on all fossil fuel power stations will
likely improve the position further over the next few years. But
will concentrated solar power beat nuclear electricity on price?
Optimistic forecasts see nuclear plants delivering power at 3.5 or
5 cents per kilowatt-hour. But these figures assume that the construction of nuclear plants can be done to the cost and timetable
set out by the contractors. The experience at the new Finnish
plant in Olkiluoto, discussed in this book's epilogue, gives us little reason for optimism.

The only real obstacle to generating most of Europe's electricity using North African solar collectors is the intimidating
scale of the TREC project. To satisfy half the U.K.'s electricity
demand (or about 180,000 gigawatt-hours a year), we would need
one thousand plants of the size of the new Andasol facility near
Granada, or two hundred plants equivalent to the planned power
station in Arizona. Achieving these numbers is perfectly feasible—there are no obvious bottlenecks involved in the world's
manufacturing industry scaling up over a decade or more to produce the reflective troughs that we need—but making it happen
rapidly will require unprecedented international cooperation. In
July 2009, a conglomerate of major banks, utility companies, and
technology businesses took the first step, linking up to form an

international consortium to build commercial North African and Middle Eastern solar power plants.

CSP could provide power for most of the world, not just Europe. The TREC project says that 90 percent of the world's population lives within 1,700 miles of a hot desert. China could get a lot of its power from the Gobi Desert, while the southwest U.S. could conceivably produce enough electricity for the whole country. Australia, with its small population and enormous resources of sun, would be able to export energy to Asia. The distances would probably be too great for electricity transmission lines, but Australia could instead use its excess power to crack water into its constituents, hydrogen and oxygen. The hydrogen, a valuable fuel, could be exported by sea in pressurized tankers.

We cannot yet know whether CSP will turn out to provide cheaper power than solar photovoltaics. But CSP has the very substantial advantage of being based on simple and easily reproducible technology. PV technology is still the exclusive preserve of a small number of very secretive companies, all understandably eager to protect their intellectual property. This doesn't improve the chances that PV will grow fast enough to decarbonize the world's electricity production any time soon. CSP has more of a following wind: many companies around the world should be able to install relatively efficient power plants. The Spanish construction companies currently leading the world have fewer technological advantages over potential competitors than First Solar or Nanosolar have in the field of photovoltaics. In the language of venture capitalists, this means that the barriers to entry for new competitors in the CSP business are relatively low and we can expect substantial competition between firms striving to drive down costs. This is not to dismiss what Abengoa and the other Spanish companies already have done, which is technically very impressive. But CSP is more scalable than solar photovoltaics, and manufacturing capacity can increase rapidly as governments and companies get excited about the technology.

Progress will be slower and more erratic than we might like, but large-scale deployment of CSP will be able to provide much of the world's energy consumption within a few decades. We should try not to put all the CSP plants in a small number of countries, which might allow them to hold their faraway customers to ransom, but otherwise there need be no restriction on where the power stations are built.

Importantly, international grids of CSP power, such as the one proposed for Europe, need to be linked with electricity-generating resources that can operate when the sun is not shining. North Africa has an average of seven hours per day of sunshine in winter and fewer than ten days on which measurable rainfall occurs. Cloudy winter days in North Africa are also likely to be the days when the wind is blowing hardest in the rest of Europe, but that relationship can't be relied upon. We can build North African plants that have heat storage, probably in the form of molten salts, but the international CSP grid will also need to be extensively linked to other sources of power, such as Scandinavian hydroelectric plants. These generating stations will have to be paid to be available at a moment's notice, ready to cascade water through their turbines if the sun ceases to shine in North Africa. Other ways of handling short-term dips in electricity supply were discussed in Chapter 1. In the worst case, we will be able to use techniques for capturing carbon dioxide and link the CSP plants to natural gas pipelines to burn fossil fuels when the sun doesn't shine.

With complementary systems such as these in place and more enlightened thinking about how to match electricity supply and demand, concentrating solar power stations could provide a huge proportion of the world's power. It's a massively exciting technology that deserves much more attention from policy-makers around the globe. With inexpensive photovoltaic panels also on the horizon, the world of solar-tech provides plenty of scope for climate and energy optimism.

3

ELECTRICITY FROM THE OCEANS

Tapping tides, waves, and currents

TIDES MOVING through narrow funnels between land masses can have enormous force. The energy contained in the moving water is almost palpable as it swirls in and out every few hours with monotonous predictability. The government of Nova Scotia announced a plan in fall 2009 to start trials with underwater turbines that can turn the power of the tides into electricity. Into the waters of the Minas Basin at the eastern extremity of the Bay of Fundy will go three different types of turbine from manufacturers in Canada and elsewhere. The goal is eventually to use the huge forces of the tide in the bay to power all of the 800,000 homes in Nova Scotia as well as parts of New Brunswick and Prince Edward Island. Many hazardous challenges remain, but the potential amounts of tidal power off both the western and eastern coasts of Canada offer a huge prize to brave developers. Across the Atlantic, even more energy is available to tap from the tides.

Go to the northeastern tip of mainland Scotland and you reach the windswept and glaciated landscape of Caithness. As the land meets the sea, the nearby Orkney Islands can some-times be seen across the 8-mile channel, often occluded by mist. Dangerous swirling currents and dramatic whirlpools make the waters a hostile place for all but the most experienced seafarers. White-crested waves mark the places with the fastest currents. The guide for ships sailing in the area says that the water in these currents can be "extraordinarily violent and confused." This nar-row sea channel between the mainland and the Orkneys is the Pentland Firth, and it has one of the fastest tidal races in the world. Twice a day, immeasurable quantities of turbulent water shift back and forth between the Atlantic Ocean and the North Sea, containing a truly huge amount of untapped energy.

The Pentland Firth is one of perhaps twenty sites around the world—from the U.S. and Canada to Australia and Indone-sia—that promise enormous potential in terms of a relatively new technology called tidal-stream power. The idea is to position turbines on the bottom of the ocean to harness the enormous electricity-generating potential of these fast-flowing tidal cur-rents. In most of the designs, these devices resemble wind turbines.

The wooden and stone windmills that dotted the hillsides of medieval Europe and Arabia were usually used to grind wheat or other cereal grains. Today's wind turbines mimic the medieval windmill. Less well known is that our medieval ancestors also built mills driven by the ebb and flow of the tides. In a technical handbook on tidal power, written in 1921, a British Army offi-cer, Major Struben, wrote that "examples of such mills exist[ed] in England, on the Breton coast of France, in America and Spain, but, as far as can be ascertained, they were only of insignificant magnitude and primitive design, and, in consequence of their intermittency, not suited to ordinary industrial uses."

This dismissive attitude to the usefulness of tidal energy was widely shared until recently. The total amount of energy in the tides across the world is not enormous, at least when compared with solar or wind power, or indeed the energy in waves. However, it is still far more than the total power needed by today's electricity grids. The energy contained in the global tides at any one moment is probably about 3,800 gigawatts, or almost twice today's worldwide electricity consumption. Most tidal energy is impossible to extract; it is found in deep oceans far from coastlines. But at a small number of places, such as the Pentland Firth or the Bay of Fundy, huge resources of energy are concentrated into narrow funnels.

Of course, tidal-stream turbines are not the only way to capture energy from the seas. Barrages are another option. These large dams harness their energy from the "range" of the tide, or the difference between its high and its low points. The barrage is built across a tidal river or estuary, and the incoming tide is allowed in through sluices. When the tide reverses, the sluices are opened, and the force of outgoing tide turns electricity turbines. We know that tidal barrages will work, as there are already commercial plants in France, Canada, and Russia.

There are also at least three marine-energy technologies that don't rely on either the range or the current of the tides. First, turbines could be positioned to collect the energy of the main ocean currents, such as the Gulf Stream. Second, wave power collectors can use the up-and-down motion of the sea as the waves pass. Finally, heat pumps can use differences between the temperature of the sea surface and the deep ocean to drive an engine, usually to generate electricity.

All of these technologies are commercially interesting, but, as this chapter shows, the power from tidal currents and ocean waves looks like the easiest to exploit and offers us the biggest potential for generating electricity.

THE POTENTIAL FOR ENERGY FROM THE SEAS

Although both the vigor and the regularity of marine energy have been obvious since humans started to sail the oceans, we have been slow to exploit their potential. Even now, only a dozen or so sites around the globe successfully generate electricity from the oceans. France built a large barrage across the River Rance to collect energy from the tides on the northern coast of Brittany over forty years ago, and a small number of other places with large ranges between the high and low tides have installed similar dams. A prototype power station in Hawaii has occasionally generated electricity from ocean temperature variations. But the general picture is of hesitant and slow progress. Only in the last few years has the pace of installation started to pick up. The first two commercial-scale turbines that capture the flow of the tide on the ocean floor have been connected to the U.K. electricity grid, and a small wave power farm has been installed off the coast of Portugal. Vancouver-based Clean Current's prototype tidal stream turbine has been successfully tested at Race Rocks off the coast of British Columbia, and a few other developers have put working machines in the water.

Why has progress been so slow? Until the recent upsurge in prices, cheap fossil fuels reduced the incentive to spend the large sums of money necessary to develop and construct devices that could profitably capture marine energy. At times of uncertainty about energy supplies in the 1970s, governments funneled some money into marine energy research programs, but the interest faded as oil started flowing freely again. Several other interesting new technologies, such as solar photovoltaic power, went through the same cycle.

None of the government-sponsored engineering trials into marine energy thirty years ago provided a definite promise of competitively priced electricity. This was particularly true when oil was only $20 a barrel. Understandably, private capital has been slow to flow into untried technologies that looked as though they would produce electricity at twice the cost of fossil fuels. Many intriguing but undeveloped ideas have been abandoned over the years as the money dried up. Few large companies listed on the stock market want to put tens of millions of dollars into risky new ventures that may never produce a commercial product. There have been some cautionary recent examples of embarrassing failures that help make investors nervous. One unlucky small company saw one of the blades break on its underwater turbine in the Hudson River near New York, while another company saw one of its devices sink to the ocean floor off the Canadian coastline.

Nevertheless, it is increasingly clear to investors that a number of other countries, such as the U.K., South Africa, Australia, Canada, Portugal, and Chile, have impressive resources of energy around their coasts. The British Isles, including Scotland, possess at least 10 percent of the world's accessible tidal energy, much of it in several extremely rapid races such as the Pentland Firth or around the island of Alderney off the coast of France. Britain also has the potential for several large barrages across estuaries and coastal bays. The tidal range of the estuary of the River Severn is one of the largest in the world, and, more importantly, the water can be relatively easily trapped behind a barrage and gradually released to drive turbines. A third advantage is that the average height of waves around the U.K. coastline is high in comparison with most other coasts.

Even though marine energy looks like a good prospect, no one should pretend that funneling government or private money into marine energy will guarantee success. Of the multitude of devices for collecting tidal energy currently in development in the world's engineering laboratories and universities, how many will survive the first few weeks of commercial trial in fast-flowing seas? Perhaps a handful. Some failed projects will be able to transfer useful knowledge to other companies, but most of the investment in marine technology is going to be wasted in fruitless endeavors.

Some of the forty designs look implausible even to a nonspecialist eye. But we are at the stage in the development of the marine renewables industry when large numbers of competitors

87

jostle for success, and we have no means of telling which are going to work. These companies need support from government if they are ever going to be able to pay the multimillion-dollar costs of designing, building, and testing sea-going machines. Of course, government-funded R&D has an almost absurdly poor financial record in most countries, and this has made governments understandably shy of handing cash over to starry-eyed engineers with poor financial skills. Nevertheless, substantial financial help will probably be needed for early-stage research, with private capital reserved for technologies that look closer to commercial launch.

Despite all the challenges, it seems entirely plausible that a small number of successful marine-energy companies will each be able to install thousands of robust turbines. In a country with long coastlines, such as Canada or the U.K., these turbines should be able to provide 20 or 30 percent of its electricity within twenty years. It took the wind power industry this length of time to move from a few lonely windmills on land owned by idealistic eccentrics to today's large generating companies aggressively bidding to install major wind farms across the world. Costs have fallen several-fold in the past fifteen years. Marine power has a similar potential to provide a substantial fraction of the electricity supply of coastal countries at prices no higher than coal or gas.

TIDAL-STREAM ENERGY

Tides are caused by the gravitational pull of the moon and, to a lesser extent, the sun. As the earth rotates, water is pulled toward the orbiting moon, resulting in small bulges in the water level. Most places experience two tides a day, though at some points on the earth's surface there is only one and at other places three or even more. The lunar cycles cause the height of the tide to vary over a period of two weeks. Spring tides—which have nothing to do with the season of the same name—are much larger than neap tides. At times of spring tides, the moon is aligned with the sun, combining the gravitational effects of the two and creating a greater pull on the oceans. At times of neap tides, the sun is perpendicular to the moon and counteracts its effect, thereby minimizing the range of the tide. Additionally, tides are larger at the spring and autumn equinoxes than during the rest of the year. So although tides are entirely predictable and reliable, the amount of energy available to extract varies substantially from week to week and by several orders of magnitude from place to place. The world's tidal energy is concentrated on a relatively small number of coasts and bays, but at these places, huge amounts of power are there for us to capture.

We cannot try to control the energy of a place like the Pentland Firth by constructing a dam across this violent and deep stretch of water, extensively used by vessels going from northern Europe to America. We need to find ways of installing devices on the sea floor that will use the power of the moving tide. Although the technical challenges are more demanding than those involved in putting a barrage across an estuary, the available energy yield from tidal races around the world is much greater.

The Pentland Firth has not been sufficiently studied for us to be sure of exactly how much energy it contains. The very ferocity of its currents makes measuring water speeds with any accuracy difficult. But most studies of the area suggest that this thin channel can generate 8 gigawatts when the tide is running at its peak. Louise Smith, a civil engineer with wide experience in road and viaduct construction around the world, was recently tempted back to this remote corner of northwestern Europe after twenty years away. Her new job is to encourage the commercial exploitation of this enormous resource of energy. She is highly optimistic about its potential and told me that some research suggests that the power in the tides of the Firth could be as much as 20 gigawatts—enough to comfortably cover the whole of London's electricity needs.

Until recently, the Pentland Firth was simply too intimidating a location for businesses even to contemplate developing machines to collect its energy. But just as the oil industry has moved into ever more inhospitable terrain as energy prices have risen, so are pioneers beginning to rise to the challenge of this most formidable of environments, hoping to harness its dangerous but enormous energy potential. In the next few years, several companies are hoping to collect energy from the tides in the Firth by planting devices on the seabed itself, either tidal turbines used singly, or ten or twenty such machines spread out in an impressive array.

The power in the tides at such places is very dense, particularly when compared with wind power. Although the speed of the wind can be several times the maximum velocity of the tide, water is about a thousand times as heavy as air. As a result, the power available is many times greater. The power in the flow of the Pentland Firth can be as much as 16 kilowatts in a vertical square foot. For comparison, the typical electricity use of a house in Europe is about half a kilowatt. So 320 European houses could be powered by the peak energy of 10 square feet of tide ripping between the open Atlantic and the North Sea. Properly located, a relatively small tidal collector could, at least in theory, produce more than the largest windmill. In a powerful tidal race, hundreds of turbines would function much as a wind farm does, collecting a good fraction of the total energy of the current.

Those backing the technology claim that the environmental impacts of a tidal farm will be extremely limited. The blades rotate slowly and are unlikely to pose much of a hazard to marine animals. The water is not trapped behind the dam (as is the case with a tidal barrage), so the ecology of the area will not be significantly affected. Perhaps these assurances are too glib—we have yet to see the effects of a full-size tidal turbine farm—but the scale of any environmental damage is probably going to be small. The best sites for tidal turbines are usually inhospitable places for fish and other creatures. The proposed

Canadian development in the Minas Basin is characterized by a seabed scoured clean by the rapid flows of water with some dunes of underwater sand and gravel.

Three or four of the world's most exploitable fast-running tides lie around Britain's coast, which looks like good news for the country. But solving the engineering problems involved in installing tidal-current generators will be extremely difficult. The force of the tides in the best locations is so enormous that machinery has to be built to extremely high specifications, while the salt and other minerals in the seawater will degrade and corrode all but the most resilient structures. The U.K. is lucky in that it already possesses one of the world's best-established offshore oil industries, one that is well practiced in providing the highest possible mechanical reliability in the face of violent seas. Since the advent of deep-sea oil production thirty years ago, we have seen improvements in designing and fabricating devices that can last for decades in unforgiving environments, and major technological advances have reduced the risk of rust, weed infestation, and water ingress.

Lunar Energy, a company that has shown some of the most exciting signs of technical and commercial progress, uses designs that come straight out of the oil business. Built in Aberdeen, the center of the U.K.'s offshore industry, Lunar's huge yellow turbine is designed to sit on the sea floor. To remain in the right place when the tides are flowing strongly, the device has to be heavy. The 1-megawatt version weighs 2,500 tons, equivalent to sixty fully laden trucks. Most of the weight is inexpensive ballast, present simply to hold the turbine in place. The machine is 80 feet long and 50 feet high. Most of the U.K.'s most powerful tidal currents occur in seas deeper than 130 feet, so the Lunar Energy turbine will sit well below the surface of the sea, minimizing any danger to shipping.

The unusual shape of the device—a long tube that narrows in the middle—helps focus the power of the tide. Water flows into the tube and then is forced to speed up as the aperture narrows.

Once past the turbine, the tube opens up again, and the water slows to the same speed as the external current. In this way, the force applied to the rotors is even greater than for a turbine that simply uses rotating blades. The rotor rotates at a sedate twenty revolutions per minute, helping to minimize wear on the moving parts. One of the many innovative features of the Lunar design is that the rotating blades do not themselves generate electricity. Their movement forces hydraulic oil through a turbine above the main chamber. All the critical components in this impressive machine can be easily removed by a boat moored above the turbine. Lunar claims that this can be conveniently done in the quiet time between tides.

One of the first tidal farms in the U.K. will be put on the seabed off the Pembrokeshire coast, on the southwestern tip of Wales. In cooperation with the huge German power generator E.ON, Lunar will install eight turbines by 2011, enough to power five thousand homes. In March 2008, Lunar announced its first export order, for a planned three-hundred-turbine tidal farm off the South Korean coast, to be completed by 2015. The machines will be built in Korea by a shipbuilding firm, avoiding the need to move these massive structures around the world. Lunar Energy gives very optimistic forecasts for the cost of electricity produced in areas of strong current, promising that once it has driven manufacturing costs down, this figure could be as low as 4 to 6 cents per kilowatt-hour, far less than carbon-intensive coal.

A large number of other firms from the British Isles are contenders for commercial success, but two businesses attract particular attention. MCT builds twin-headed windmill-like turbines. The structure is supported on a single steel pile that has been driven into the seabed. Unlike the Lunar device, MCT's machine can be raised above the sea surface for maintenance. The disadvantage of this design is that a portion of the structure is always above the sea surface, meaning that it is potentially more of a risk to marine traffic. The first large-scale installation of the MCT turbines was in Strangford Lough in Northern Ireland in March 2007, six months late because of an agonizing wait for access to one of the small number of specialized ships designed for deep-water installation projects. (The U.K.'s potentially huge offshore wind industry will probably also be held up by this crippling worldwide shortage of vessels that can carry out work on installations of this sort.) A plan for a seven-turbine farm using MCT devices off the coast of North Wales is backed by RWE, the second-largest German utility. One of the crucial reasons this location was chosen, apart from its tidal speeds, was the availability of a nearby connection to the electricity transmission system. One of the problems with tidal energy around the world is going to be connecting the turbines to the power grid. The best tidal locations tend to be far from high-voltage transmission lines and to use the full force of the Pentland Firth, the U.K. will need a new offshore undersea cable running down the east coast to London and on to the rest of Europe, using the same HVDC technology that will bring Saharan solar power to Germany and other countries, as described in Chapter 2.

The third high-profile competitor for the wide-open market for tidal power generators is the Irish firm OpenHydro. This company produces a striking O-shaped device, consisting of a central rotor that spins inside an outer ring, generating electricity as it moves. This extremely elegant system requires only one moving part, and, as with the Lunar Energy turbine, the whole structure sits well below the surface, posing little risk to shipping. The first commercial OpenHydro devices will be installed in the tidal races in Canada's Bay of Fundy or in turbulent seas off Alderney. This small island, lying near the coast of France, but with close ties to Britain, has tidal flows that rival the Pentland Firth for their concentrated power. One estimate suggests that peak flows at Alderney would be equivalent to 5 percent or more of U.K. electricity use.

Canada's Bay of Fundy is a similarly important location—it is 180 miles long and 60 miles wide and probably has the largest tidal range in the world. On rare occasions, high spring tides can be almost 65 feet higher than low tide. The twice-daily water flows in and out of this channel are greater than all the rivers and streams in the world combined, making it an obvious target for power generation. One previous attempt to exploit a very small fraction of the water's power was made in 1984 by constructing a tidal barrage across a river flowing into the bay. The interest in tidal power has increased in recent years, and power utilities have begun to plan experiments to use tidal-stream generators as well as new tidal barrages in this enormous bay. As in several other parts of the world, the most promising of the many different models of tidal generator are going to be tested for effectiveness and robustness in the next couple of years.

On the other side of Canada, British Columbia has identified ninety separate sites that have enough tidal current to make extracting energy worthwhile. Most are close to Vancouver, the main center of electricity demand, meaning that exploitation there is particularly attractive. These sites could produce 4 gigawatts, a sizeable fraction of the total demand from the whole of British Columbia. As in the British Isles, the local availability of tidal energy has spawned the early beginnings of an industry trying to commercialize designs for tidal power. Clean Current, a Vancouver firm, has installed an experimental turbine of the same style as OpenHydro's in the waters of British Columbia and is involved in the bigger trials in the Bay of Fundy.

The tidal-stream generator designed by Australian firm Bio-Power is based on copying what the company calls "the highly efficient propulsion of Thunniform mode swimming species, such as shark, tuna, and mackerel." The first commercial prototype of this device will be installed off Flinders Island, Tasmania, in 2010, supplementing the diesel generators and wind turbines that provide the small community's electricity. The company hopes to have a 250-kilowatt machine for sale by 2011.

Predicting the success or failure of technologies at an early stage in their development is not a game sensible people play. In the case of tidal currents, forecasting is even more difficult than usual because of the multitude of different turbines in development. It is impossible to predict which will succeed. But will one or more companies manage to develop a turbine that reliably and inexpensively generates electricity at significantly less than 7 cents per kilowatt-hour, making it broadly competitive with fossil fuels? I think the answer is almost certainly yes. We do not have to challenge the laws of physics or those of thermodynamics as we have to do with some other technologies. The crucial problems are those of mechanical engineering and are thus more susceptible to eventual solution, probably by continuing to improve strength and robustness.

BARRAGES THAT USE THE RANGE OF THE TIDES

In places such as the Bay of Fundy and northward into Labrador, where the difference between high-tide and low-tide levels is large, the most obvious way of capturing energy may be to build a barrage rather than installing tidal-stream machines on the sea floor. Barrages usually work by letting the incoming tide flow freely though an embankment. When high tide is reached, and a large body of water is sitting behind the dam, the gates that allowed the water to flow through are shut. From this point on, the barrage works in exactly the same way as a hydroelectric power plant. As the tide falls, reducing the water level outside the barrage, a height difference develops between the water levels behind and in front of the barrage. Water can then be allowed to flow through the dam toward the lower level, turning turbines as it goes and generating electricity. The best sites for barrages are likely to be enclosed bays or large estuaries.

Tidal barrages could, in theory, generate electricity when the water moves through the turbines in both directions, inward as well as outward. But in the small number of barrages currently operating around the world, capturing energy at both phases of

95

the tide has not proved easy, and they can only produce electricity for about half the day. The amount of power generated will peak a couple of hours after the high tide has passed and will then fall away.

Spring tides will provide far more power at tidal barrages than neap tides because the range is typically twice as large. So the amount of power available to a national electricity system will vary in a regular cycle. Very usefully, however, the peaks of spring tides always occur at the same time of day at any particular location. So, for example, we know that most power from the proposed barrage across the River Severn in England will come between 1 and 3 PM. But other places around a coastline will have their peak at different times. This means that a portfolio of tidal barrages and turbine farms spread along a long coastline will potentially provide very stable levels of power throughout the day. Tidal power can thus avoid some, but not all, of the intermittency associated with wind and solar power.

We will be able to forecast reasonably accurately how much power will be delivered every minute, years in advance. That said, as with tidal current power, the amount generated will vary through the seasons, with the tides at equinox being more powerful than at other times. The weather will also slightly affect power generation, since strong winds influence the height of tides, though this effect will likely be quite minor. All told, a wide spread of tidal generators will be almost as useful to the operators of an electricity grid as a coal-fired power station.

The River Rance dam has been successfully generating a peak 240 megawatts (enough to meet the electricity needs of almost half a million French homes) for several decades. The best-known scheme for a much larger tidal barrage is the one proposed for the Severn, the major river draining much of Wales and parts of western England. The Severn has a huge tidal range—probably second only to a few places on the east coast of Canada—because the tide is funneled up a sharply narrowing estuary. A single dam across the river, about 10 miles long and costing about $25 billion, would generate roughly 5 percent of the U.K.'s electricity consumption. It is not all good news, however: a U.K. government body has estimated that if constructed by private capital, the project would produce electricity at a cost of over 15 cents per kilowatt-hour, much higher than many other types of renewable technology. Other dams around the U.K. coastline could provide another few percent of the country's electricity but possibly at an even higher cost.

In addition, tidal barrages have some unfortunate side-effects. They change the ecology of the areas behind the dams because they reduce the range of the tide and impede the movement of silt. The Severn barrage would also cause high levels of fish mortality, since many of the unfortunate creatures would be sliced into fillets as they passed through the turbines, and would probably disrupt the colonies of birds that feed in the estuary. However, the high price of most tidal barrages is what really upsets their viability. While the cost of, say, mechanical turbines falls as we construct more and more of them, the same will not likely be true of building huge one-off concrete dams many miles long.

Another important reason for skepticism about the value of tidal dams is the relatively small number of locations—possibly well under a hundred across the globe—where significant amounts of power are available. For all these reasons, tidal barrages remain a less exciting technology than tidal-stream turbines, which promise fewer side-effects and a greater potential for plentiful, inexpensive power.

97

POWER FROM THE WAVES

Tidal energy is generated by the tug of the moon and sun, but wave power comes from winds lashing the oceans. Wave power is therefore a much-mutated form of solar energy. The sun heats the air, and temperature differentials between areas produce wind, which then generates waves when it passes over the sea. Less than 1 percent of solar energy becomes wind, and a very small percentage of wind power is then transformed into waves. These facts would appear to make collecting wave energy less attractive than harnessing solar or wind power. But in some ways, waves are better. The crest of the waves in the open Atlantic seas off the coast of Portugal and the Pacific seas off Washington State might contain as much as 6.5 kilowatts in each square foot of sea. If captured, the power from this tiny area would be enough to heat a drafty house in the depths of a cold winter or provide the typical electricity needs of almost seven homes. By contrast, a square foot of hot tropical desert might only receive a tenth of a kilowatt of solar power even when the sun is high in the sky.

Tidal power is concentrated in a small number of areas, but wave power is widely spread around the globe. And although the best tidal sites may contain more energy, wave power is available for hundreds of miles along straight shorelines. The best areas tend to be in temperate zones, where western coasts are exposed to the prevailing winds and where frequent strong storms whip up high waves. Very high levels of available energy have been measured off coastlines as diverse as southern Chile, western Australia, Portugal, and South Africa.

Nevertheless, we will never find collecting more than a small fraction of all wave energy worthwhile. The largest waves are found in open seas, hundreds of miles from coastlines. We might be able to install energy collectors there, but transporting the energy to the nearest electricity grid would be costly. Moreover, an important design consideration affects almost all wave collectors. The stormiest seas contain truly awesome amounts of power. If we tried to collect this energy, the devices would have to be extraordinarily strong, able to resist ferocious forces, occasionally exceeding what is seen in even the most powerful tidal currents. To withstand these forces, the machines would have to be so robust that the cost of manufacturing them would be crippling. Much like wind turbines, which close down in gales, most of the wave collectors currently being tried out around the world don't attempt to operate in storms, and thus they don't harvest energy when the waves are at their most energetic. In this and other ways, wave collectors tend to be designed for survival and not for maximum energy output. The Australian company BioPower, whose fish-fin tidal-stream generator I mentioned above, has also developed a wave power device. Its prototype, which mimics the actions of water weeds in turbulent water, does not try to resist the force of the waves but simply lies flat on the ocean floor when the energy of the seas is too violent.

Even if they harness a few percent of the passing energy, wave devices are capable of fulfilling a large fraction of the total energy needs of many countries. For example, the British wave industry trade body claims that the total amount of accessible wave energy in U.K. waters is about twice the country's total electricity use. (The leading British wave power developer, Pelamis, gives a similar figure but qualifies it by saying that only a portion might be economically recoverable.) The industry trade association also quotes analysis that suggests the worldwide availability of electric power from the waves might be as much as four times current global electricity use. Other sources claim even higher figures. Whatever the correct number, wave power should be able to service a large fraction of our needs for electric power.

A host of different wave energy approaches are jostling for the attention of banks, governments, and electricity companies. One industry website reports that a search of the patent literature throws up more than a thousand proposals for wave energy collectors. But, as with tidal energy, we can be sure that most of these collectors, if built, would be in pieces on the sea floor within a few hours of the start of a severe storm. Indeed, only a handful of prototype wave devices have ever exported power consistently. Nevertheless, increasing interest in renewable energy, higher levels of government funding, rising electricity prices, and better

99

construction techniques have combined to make wave power an exciting area to watch.

The first commercial wave farm in the world lies 3 miles off the coast in northern Portugal. At this distance from the shore, waves are much more powerful than they are nearer the coast, since their energy dissipates as they run up to the beach. Swell from the Atlantic is captured by three Pelamis machines, made in Scotland but hooked up to the Portuguese grid. Generating about 750 kilowatts each when the wave conditions are right, these long, red articulated cylinders are the fruit of over thirty years' work. The story begins in the engineering workshops of the University of Edinburgh. In 1974, just as the world was waking up to the potential volatility of energy prices after the Arab oil embargo, Professor Stephen Salter came up with the idea of a device that productively absorbed energy from the waves. Even then, it was obvious that a wave collector could potentially capture 80 percent or more of the power of a wave: this figure is far higher than a photovoltaic device converting the energy of the sun will ever likely achieve and a higher percentage than the theoretical maximum that a wind turbine can achieve.

Photographs from the period show engineering students in the 1970s employing arrays of analog electronics to control the waves in a tank and to measure the energy captured by what became known as Salter's Duck. The 500-foot-long red Pelamis machines that float semi-submerged in the powerful waves off north Portugal are the very indirect descendants of this work, benefiting from many generations of prototypes and thousands of hours of testing in the wave research center in the Orkneys off northern Scotland. These huge structures, weighing 750 tons each when filled with ballast, are the nearest thing the world has to proof that wave energy can be profitably extracted. They've had mechanical problems—and the majority owner of the wave farm fell victim to the 2008 financial crisis—but the technology has proved itself, and German multinational utility E.ON has placed an order for the next generation of Pelamis devices to be installed off Scotland.

The Pelamis is composed of four cylindrical segments, each joined to the next by a flexible link that contains the power-generation module. The whole snake-like device is loosely moored on the sea floor and aligns itself automatically at 90 degrees to the prevailing wave direction. As the wave passes along the Pelamis, individual segments rise and fall. This motion causes the joint between the cylinders to flex, pushing hydraulic rams, which pump oil under pressure. This pressure is converted into electrical energy via a turbine. The electricity from multiple machines is then combined and sent onshore. As with many of the other interesting wave capture devices, the machines are designed and built by people with experience in constructing massive steel structures for the Scottish offshore oil industry. They are engineered to last decades, but if they need maintenance, these steel sea snakes can be unhooked and easily towed into port for repair.

Portugal has excellent wave power, and its shores are an obvious choice for the first attempt to put multiple machines into the water. Atlantic westerlies whip up the seas along the coastline, generating high and relatively reliable waves. In September 2009, Energias de Portugal, one of Europe's major electricity utilities, bought out the existing owner, and all is now set for the first three Pelamis machines to be followed by twenty or thirty others, spaced over a third of a square mile of ocean. The company commercializing the Pelamis says that the technology could generate over 30 megawatts in this area. This is the same amount of power as six of the very biggest offshore wind turbines would generate, and they would probably require a much larger area of sea.

The price initially paid for the three Pelamis generators was about $10 million, a substantial premium over the cost of a wind farm of the same power. But wave power is likely to be important to Portugal, which has no significant fossil fuel resources, and this investment is small in the context of the size of the opportunity. Wind generation is growing fast but will eventually slow as the country runs out of good onshore locations. One prediction is that waves will produce almost a third of the country's electricity by 2050. The electricity company that constructed the Pelamis wave farm will want to capitalize on its early experience and capture itself a substantial fraction of this market.

The second crucial reason Portugal is the first place to install a working wave farm is that the electricity companies there are required to pay a higher price for power generated from the sea. Government regulations mean that the owners of the Pelamis machines will get about 35 cents per kilowatt-hour, several times the standard wholesale price for electricity. When the farm is working at its peak rate, it will be earning nearly $700 an hour. Even at half capacity, the three machines will produce over $3 million worth of electricity in a year, providing a reasonable return on the initial investment. But without the high prices available as feed-in tariffs, the Pelamis would be far too expensive.

But once the costs of the Pelamis fall, or the price paid for wave-generated electricity rises, the rush for wave power will start across the west coasts of Canada and the northwest U.S. and many other places around the world. An area of 400 square miles could potentially provide over half of California's total need for power. Another suggestion is to create a long strip of wave farms a mile wide around the exposed western coasts of England, Wales, and Scotland, taking the electricity onshore at points where the local grid connections are strong. Even though the scale of such a wave farm would be enormous, it would be unlikely to change local ecology much. Although the Pelamis is very long, it sits only about half a dozen feet deep in the water and seems to cause minimal disruption to flows of fish and sea mammals. Installing banks of such devices intermingled with the huge offshore wind farms that are very likely to be built in similar places would seem logical.

Other countries with substantial wave resources could be significant beneficiaries once wave collectors have reached commercial viability. Take South Africa, where electricity demand is growing fast but production has failed to keep up, with blackouts repeatedly shutting the gold mines and costing the country billions in lost exports. Wave farms at South Africa's southern tip could provide at least 20 percent of the country's power needs, reducing the shortfall without raising carbon emissions. In fact, almost all countries outside the tropics with long west-facing open coastlines could generate a large fraction of their power by deploying wave collection devices.

Pelamis's sponsors think that their product has a good chance of eventually producing electricity at prices equivalent to or lower than those of fossil fuels. Independent estimates of future costs are falling, too, as analysts see that large-scale manufacturing will improve prices. But there have been significant delays in commissioning the Pelamis machines in Portugal, and we cannot know yet whether this particular approach to collecting wave power will be the one that gets adopted around the world. Investors have already put $70 million into the company developing the Pelamis, a huge sum by the standards of the European renewables industry, though trivial in the context of the money that has gone into solar photovoltaic or cellulosic ethanol plants in the U.S. Investors must be hoping that the long wait before their machines can be used is almost over.

What if the Pelamis turns out not to work or to be too expensive to be widely used? What is the next most plausible type of wave power collector? Most attention is focused on devices that bob up and down in the water like fishermen's floats. This heaving motion drives a piston that moves a pressurized column of fluid inside the device, which in turn provides the power to drive a turbine. Small buoys in harbors often use similar technologies—but on a much smaller scale—to power the light that alerts ships to their presence. Canada's Finavera and New Jersey's Ocean Power Technologies are among the companies hoping to apply this technique for large-scale generation. Growing interest in harnessing the power of the waves off the west coast of

103

the U.S. has begun to help these businesses gain attention, but their technology is still at an early stage. In late 2007, one of Finavera's prototype buoys sank after a two-month ocean trial. Reports at the time suggested that seawater had gotten into part of the mechanism after a pump had failed. Despite this setback, Finavera went on to win a contract to build wave devices off the coast of Washington State and to get support from California's largest power-generation company for another project.

A third type of wave generator also offers some potential. Many harbors and beaches have a breakwater, a wall in the sea designed to minimize waves and create stiller waters. Breakwaters can be designed to catch some of the water from the waves as they crash into the stones or concrete. Once trapped, the water escapes downward through a shaft and can drive a turbine, much like a hydroelectric power plant. The mechanical challenges of making such a device are minor since the breakwater itself has taken most of the force out of the waves, but the consequence is that relatively little energy remains to be used. These devices could be used in coastal protection schemes in the future but are unlikely ever to generate a substantial fraction of world electricity need.

More generally, though, wave power offers enormous potential. The mechanical engineering challenges are probably as substantial as those for tidal current turbines, but they're clearly solvable within a few years. Indeed, they're almost trivial compared with those faced by some parts of the offshore oil industry. All that's needed is support from governments—a high guaranteed electricity price and continued funding of university research—and help from large energy companies and private investors. With that support in place, a Pelamis-like device will eventually produce cheap electricity on a large scale around the world's western coasts.

THE GULF STREAM AND OTHER OCEAN CURRENTS

The Gulf Stream isn't a tide, although it has some of the same useful features. It is a continuous flow of water moving from the

Caribbean to the northern Atlantic, where the current sinks and returns south. This circular motion is part of the system of worldwide ocean currents, driven by the winds and by differences in water density around the globe.

The easiest place to exploit these currents lies off the coast of Florida, where the majority of the kinetic energy of the Gulf Stream is funneled into a zone just 60 miles wide. Here, turbines that resemble the MCT tidal collector at Strangford Lough in Northern Ireland could be used to extract some of the power for conversion into electricity. At one point, the Gulf Stream runs only about 15 miles off the Florida coast, meaning that connecting to the grid shouldn't be a major problem.

At about 5 miles per hour, the Gulf Stream moves much more slowly than the tidal races in, say, the Pentland Firth. This fact is significant, because the energy in a moving stream of water (as with the wind) goes up exponentially with speed—a stream moving at 10 miles per hour has eight times the force of a water of 5 miles per hour. So the relatively sedate Gulf Stream will never match the energy potential of the Bay of Fundy if we put only a few turbines in the sea. But the Gulf Stream is wide enough to allow us to install thousands of slow-moving devices; it also has the huge advantage that the speed is reasonably consistent throughout the year, unlike with tides. One researcher has calculated that the Gulf Stream ought to be able to provide a third of Florida's electricity needs.

Nobody doubts that building effective turbines in the Gulf Stream is possible, since the waters are much less fierce than those off northern Scotland. The calmer water speeds mean that the turbine blades can be much bigger and more like windmills. The issue is money: given the low speed of the current, will it be possible to build an underwater mill that can cover its costs? It all depends on the price of electricity in Florida. Other ocean currents, such as the relatively fast-flowing Kuroshio off Taiwan, could also be used for generating electricity in places where the worldwide oceanic conveyor belt, as it is sometimes known, comes close to populated coasts. Unfortunately, not many places meet these criteria.

What about the wider environmental effects of slowing ocean currents by placing turbines in their way? Should people be worried in Britain, Norway, and the rest of northwestern Europe, where potentially bitter winters are kept several degrees warmer by the northern extension of the Gulf Stream? I think not, because proponents of Gulf Stream energy collection only intend to capture about a thousandth of the kinetic energy of the current.

OTEC

The final potential source of energy in the oceans is altogether different from the ones discussed so far. The idea is to exploit not the movement of water but the difference in temperature between the warm surface waters and the colder depths. In parts of the central tropics, this temperature gradient is more than 20 degrees Celsius (36 degrees Fahrenheit), which in theory means these waters might warrant the installation of power stations based on the principles of the heat pump. The approach is known as "ocean thermal energy conversion," or OTEC.

In a closed-circuit OTEC plant, the hot surface water passes through a heat exchanger, causing a low-boiling-point liquid, such as pressurized ammonia, to turn into gas. The expanding gas drives a turbine, thereby generating electricity, before exchanging heat with the cold water and condensing back into a liquid. And so on.

As with many other technologies discussed in this book, the basic idea is not new: scientists originally worked out how to turn a temperature difference into electricity over a hundred years ago. Several attempts to build a working generator have been made over the decades, but the availability of cheap fossil fuels has always disrupted the experiments. Government research and development money dries up a few months after the price of oil begins one of its periodic slips.

Even with higher fossil fuel prices and fears about climate change, however, OTEC is unlikely to prove a key technology in the coming decade. The main problem is that the ocean's surface temperature cannot rise above about 31°C (88°F). Higher than this level, the energy lost through the evaporation of water cancels out the energy gained from the sun. Since very few places near coasts have deep water at less than 4°C (39°F), the maximum temperature gradient is about 27 degrees Celsius (49 degrees Fahrenheit), which is too small to create a very efficient heat engine. The best system might only capture a few percent of the energy gradient and will require significant amounts of power to run itself. Although the technology may well yield a small positive energy balance, few but the band of enthusiasts backing the technology have much confidence that it will ever produce economical electric power, particularly because this technology will only work in the tropics, where concentrated solar energy may become a cheap alternative.

Oceans as a whole, though, possess enormous untapped potential for low-carbon energy creation. Although it is still early days, I suspect that the best bets are wave collectors using the same principles as the Pelamis devices operating off Portugal and huge steel tidal-current collectors such as those produced by Clean Current and MCT. Manufacturing such devices is relatively simple, and producers are unlikely to face any pressing shortages of raw materials. Once the devices are proven, there needn't be any huge delay before they're installed in their hundreds of thousands in oceans around the world.

COMBINED HEAT AND POWER

Fuel cells and district heating

MOST ELECTRICITY generation today is inherently wasteful. Old coal-fired plants turn only about a third of the energy in their fuel into electricity. Even the best new gas plants struggle to reach 60 percent. The rest of the energy becomes heat, which is treated as a waste product and frittered away in a cooling tower. Just down the road, thousands of people create yet more carbon emissions by burning gas to heat their homes. It doesn't really add up—environmentally or economically. A rational system would use the heat created in power stations to replace central heating boilers.

We have two obvious ways of preventing this enormous waste of heat. We can switch to small power stations close to homes or offices and pipe the "waste" heat where it is needed. Or, on an even smaller scale, we can install microgenerators in our houses and places of work, making electricity precisely where and when we need it, and using the accompanying heat for room and water heating (or even for cooling, via a clever process known as adsorption chilling. These two very different approaches both go by the broad name of "combined heat and power," usually abbreviated to CHP.

Everybody agrees that CHP is a good idea: the towers of steam billowing out of large power stations are an increasingly obvious symbol of our profligacy with the world's scarce energy resources. But despite the attractiveness of the basic concept, CHP has struggled to grow in most countries. Using the waste heat from a small district electricity plant requires a network of underground pipes to take hot water to local homes and offices, an up-front cost that has provided a barrier during an era of cheap fossil fuels. One entrepreneur I talked to told me that it would cost $73 a foot to install the insulated pipes for a heating scheme in a European urban center. Tiny heat-and-power units in individual buildings get around this problem, but unfortunately, electricity generation on such a small scale has been very inefficient up until now, which means limited savings of carbon dioxide. Although CHP makes economic sense for some industrial processes in which the factory needs both heat and electricity, the costs have meant that adoption of the technology has not been particularly fast in recent years.

Thankfully, the obstacles to more widespread use of CHP are gradually disappearing. High fuel costs are making the installation costs of district heating pipes look more reasonable, and the efficiency of micro-CHP units is increasing. This chapter focuses on the two most interesting prospects for taking CHP forward: fuel cells powered by hydrogen created from renewable sources for individual buildings, and district plants powered by wood and other biomass, again with a low carbon cost. Both approaches offer heat and electricity with minimal waste of heat and no use of fossil fuels.

FUEL CELLS

A fuel cell is effectively a battery. One side of the cell has a positive charge and the other has a negative charge. The two sides are separated by a semi-porous material called an electrolyte that allows electrically charged atoms to flow through. When an external wire connects the two sides, a current will flow around the entire circuit, just as with a battery. But unlike a standard disposable battery, we can top up the electric power of a fuel cell by adding more fuel, usually hydrogen on one side and oxygen on the other. For as long as there is fuel in the cell, a chemical reaction strips the positive charge from oxygen atoms and the negative charge from hydrogen, resulting in a reliable and consistent flow of electricity.

Getting oxygen into the cell is easy. The gas makes up a fifth of the atmosphere, so fuel cells simply feed ordinary air to the positive pole of the battery. Hydrogen is trickier. The pure form of this very light gas does not occur freely at ground level because it quickly escapes upward to the ozone layer, where it then reacts to form water and oxygen. (Loose hydrogen is therefore an ozone-depleting chemical.)

One option for producing pure hydrogen is to create it on an industrial scale by splitting water or hydrocarbons and storing the resulting gas in tanks that can be hooked up to fuel cells, where the gas will be efficiently and safely used. Despite its reputation, hydrogen is not particularly flammable or explosive.

Alternatively, the hydrogen can be made in the fuel cell itself. Perhaps the simplest way to do this is to use methane, the main component of natural gas, as the fuel. When heated to a very high temperature in the presence of steam, methane separates into its constituent elements: hydrogen and carbon. The carbon atoms combine with the oxygen in the water molecules in the steam to create carbon monoxide. (Usually known as "steam reforming," this is the same process as discussed in Chapter 8, on carbon capture and storage.) This process leaves pure hydrogen gas. Inside the fuel cell, the hydrogen atoms then separate into their constituent parts: protons and electrons. The electrons, unable to travel through the innards of the cell, flow around the external circuit, providing electricity. The oxygen and hydrogen atoms eventually combine to form water, which is one of the two waste products of the process. The other is carbon dioxide, made from the oxygen and the carbon monoxide produced from the steam reforming of the original methane feedstock.

The voltage in an individual fuel cell is small—as little as 1 volt, which is less than that of an A A battery. So to make useful cells capable of driving large machines, many tiny cells are wired up together into a power pack. This means that fuel cells can vary in size from small devices for powering laptop computers or mobile phones to machines the size of several shipping containers capable of providing large buildings with all their electric power.

Like several other technologies in this book, fuel cells have been around for a long time but still have not achieved their full potential. The first cells were created in about 1842 by Sir William Grove, a scientist and lawyer originally from Swansea in South Wales. Intermittent attempts to create a commercial use for the technology followed, and in the 1950s, several businesses tried to develop fuel cells to power vehicles or satellites. As with solar photovoltaic cells, however, the early promise shown in space missions proved difficult to translate into wider commercial success. For at least the last twenty years, several dogged manufacturers, such as Ballard Power in Canada, have been trying to build fuel cells that successfully compete with other energy sources, usually focusing on vehicles such as city buses. Progress has been painfully slow, for although the technology is well understood, delivering a powerful fuel cell at a price that can compete with a standard internal combustion engine has been difficult. Because some types of fuel cell, including Ballard's, use catalysts made from rare materials such as platinum, rapid upward shifts in metals prices have also impeded their development.

THE MOST PROMISING TYPE OF FUEL CELL

Several types of fuel cell are in production today, and they vary in important ways. Some need pure hydrogen as a fuel, while others create their own from gases or liquids. Some operate at low temperatures, while others work at the high temperatures necessary to split fuels into hydrogen and other chemical elements. Electrical efficiencies vary dramatically between the various types, but the best cells can now turn more than half of the usable energy in natural gas into electricity.

Ceramic Fuel Cells in Melbourne, Australia, is one of several businesses making good progress in constructing fuel cells for generating electricity and heat on a domestic scale. The company's cells are fueled by the conventional domestic gas supply (which is largely methane) and use one of the most promising fuel cell technologies, usually called the "solid-oxide" approach. This description refers to the substance, made from the ceramic-like compound zirconia, that functions as the electrically porous center of the cell. This technology uses very high temperatures, about 700°C (1,290°F) and takes some time to get started, but it does not require the extremely expensive platinum catalysts used in other types of fuel cell.

Ceramic Fuel Cells' solid-oxide home power plants generate up to 2 kilowatts of electricity. The fuel cell system is almost 60 percent efficient, meaning that it generates over 2 kilowatt-hours of electricity from natural gas that would produce 4 kilowatt-hours of heat if it were simply burnt. However, most of the rest of the energy produced by the fuel cell can be used for heating, meaning that the device may be able to capture 85 percent of the total energy of the gas and put it to use in the home.

By contrast, a new, ultra-efficient, large-scale gas power station can also turn about 60 percent of the energy value of gas into electricity but will suffer 5 to 10 percent transmission loss in getting the electricity from the power station to users in homes and offices. In other words, the best fossil fuel plants are no better at turning gas into usable home electricity than one of Ceramic Fuel Cells' domestic-scale units. And when you factor in the useful heat generated, the domestic unit looks better overall.

In principle, solid-oxide fuel cells are very effective

competitors to mainstream power generation. But, as the long-suffering investors in fuel cell companies are eager to explain, the entrepreneurs still have many hurdles to overcome. For one thing, there's size. Domestic fuel cells are still bigger than most natural gas furnaces. Even after a major effort to reduce the size of its domestic cell, Ceramic Fuel Cells' prototype is still the same size as a washing machine.

Then there's cost. Most of the companies in the fuel cell industry are coy about the price of their units, and Ceramic is no exception. We can safely assume that its product is still substantially more expensive than conventional central heating apparatus, although manufacturers are all promising continued sharp cost reductions. To get mass acceptance of the technology, solid-oxide fuel cells need to be priced to deliver electricity competitively. In the case of fuel cells, the industry thinks that means getting to below $2 per watt of continuous electricity output, or $4,000 for a 2-kilowatt cell.

Perhaps just as importantly, manufacturers like Ceramic Fuel Cells have been struggling to improve the lifetime of some of the components in the cell. Most solid-oxide fuel cells currently last only three or four years before some components will need to be replaced or refurbished. Customers would typically expect at least a ten-year operating life, and perhaps even longer, for a domestic natural gas boiler.

Another issue is that the attractiveness of domestic fuel cells will significantly depend on the prices that homeowners can get for electricity they feed into the grid. The first home CHP units were installed ten or more years ago. They used conventional internal combustion engines, not fuel cells, to generate heat from gas. The electricity generated came as a bonus. When the heat was not needed, such as in the summer or when the family was away, these CHP boilers were turned off, and the home took its electricity from the grid. Ceramic Fuel Cells has made many interesting innovations, but its key insight is that because its product can deliver an electricity output of well over 50 percent of the energy value of the gas used, it makes sense to operate the unit twenty-four hours a day and export excess electricity to the grid. The unit could therefore be kept on even when the home is unoccupied. If the house does not need the associated heat, it can simply be exhausted to the open air, as at a power station. Some European countries, including Germany and France, have regulations that discourage the use of fuel cells when the heat is not actually being used, creating an obstacle to rapid adoption of these innovative devices.

The typical home will only use the maximum electricity output for a small portion of the day. (The electricity use in a North American home is about a kilowatt, averaged over the twenty-four-hour day.) So the Ceramic Fuel Cell unit would satisfy the demands of the household almost all the time and would export electricity from the house into the local grid. The value of the unit to the homeowner thus depends largely on how much he or she gets paid for this exported power. In places with a "net metering" requirement, including some American states and Canadian provinces, the electricity utility is obligated to pay the same amount for exported electricity as it charges for household consumption. This scheme provides a good deal for the fuel cell owner, who makes money running the cell all the time and exporting electricity into the local grid. At current gas and power prices, the cell's owner might make a profit of $1,000 a year from electricity sales, even if all the heat is evacuated to the outside air.

But the heat needn't be wasted. If it works as promised, the fuel cell will provide all the heat and hot water needed for a very well-insulated house during the winter, even in quite cold countries. A big Passivhaus (see the next chapter) could be heated by a single fuel cell. In less energy-efficient houses, an integrated high-efficiency condensing boiler would be needed to supplement the heat. During the summer, the fuel cell would simply heat hot water for washing, venting the spare heat to the outside.

As equipment costs come down, homeowners will find buying fuel cells and running them as micropower stations increasingly attractive.

However, Ceramic Fuel Cells does not actually plan to sell its fuel cells to homeowners. It will provide them to the electricity utilities, which will then lease them to individual homes. Although the first units for commercial sale will not be produced until summer 2010 or even later, the company has pre-orders for fifty thousand units from a Dutch utility and other orders from Germany. Relatively high local energy prices mean the early market for its fuel cells will probably largely be in Europe, so Ceramic is building its first factory in Heinsberg, Germany. The critical ceramic component, zirconia, that sits between the positive and negative poles of the cell is made in a specialized plant in northern England.

OTHER TYPES OF FUEL CELLS

While Ceramic Fuel Cells is focusing on homes, other manufacturers aim to address the market for larger machines for apartment blocks, offices, supermarkets, and hospitals. U.S. companies such as FuelCell Energy in Connecticut are already delivering units that produce more than 1 megawatt, enough to power an office block or a small shopping mall. Although Fuel-Cell Energy's products are not currently based on the relatively new solid-oxide technology, they are efficient, reliable, and attractive to utility customers. Like their smaller cousins, these large units can transform about 50 percent of the usable chemical power of a fuel into electric power as well as generate large amounts of usable heat, either for keeping the building warm or cooling it via an adsorption chiller in summer. A 1-megawatt device will take up relatively little space and can sit adjacent to the building. The fuel cells have few safety issues, and, unlike the backup diesel generators frequently used by hospitals and other large buildings, they cause no local air pollution. FuelCell Energy uses molten carbonate technology for its cells, although it is hedging its risks by also actively participating in the U.S. government's research program to improve the solid-oxide approach. (One-megawatt solid-oxide fuel cells are likely to be available within the year from such companies as Rolls-Royce.)

FuelCell Energy's most important customer is the leading independent Korean electricity company POSCO Power, which has ordered dozens of megawatts of capacity for delivery over the next few years, including some individual units as large as 2.4 megawatts. An installation of this size could produce about as much electricity each year as two large wind turbines in a windy location but only takes up a fraction of the space.

Korea is poor in indigenous sources of energy, and its government is actively backing the use of fuel cells for electricity generation by guaranteeing high prices for their electricity output, similar to the program of feed-in tariffs widely used in countries such as Germany. The feed-in rates for fuel cell electricity in Korea are currently twice or three times the typical wholesale price of power. This substantial price incentive (which will gently decline over the next few years as the technology matures) has meant that many of the early orders for large-scale fuel cells have come from this country. The total installation is, of course, still only a small fraction of the output of a large coal or gas power station. At about $3,000 per kilowatt of power output, FuelCell Energy's plants are still expensive, though they compare well, for example, with the price of wind energy just a few years ago.

Japan has used a different approach, focusing on domestic rather than commercial fuel cells. The Japanese government has actively supported the installation of smaller cells by subsidy that rebates a large part of the unit's cost. In the first few years of the program, each installation received a grant of tens of thousands of dollars. The amount is decreasing annually as the cost of the cells falls. Progress has been much slower than the optimists predicted: in 2003, Japan announced what now looks like an absurdly ambitious target of generating 4.5 percent of all its electric power from fuel cells by 2010. But the generous subsidy scheme has helped to start active competition in Japan between the proponents of the different fuel cell technologies. Indeed, this chapter's prediction that solid-oxide fuel cells will win the day is almost certain to be tested first in Japan. Ceramic Fuel Cells' units will be imported into Japan by its local partner, a large central heating boiler manufacturer, while the electronics giant Panasonic is putting its efforts into a different technology, similar to that of Canada's Ballard Power. Ballard itself is offering its low-temperature and well-established proton-exchange membrane approach in partnership with a Japanese business.

Panasonic expects that by about 2012, Japanese homeowners will find that it makes financial sense to purchase its fuel cells without government subsidy. By 2015, it is forecasting sales of 300,000 domestic units, a very large number but still less than 10 percent of all heating systems installed annually in the country. The manufacturing cost is still predicted to be high, at almost $5,000 per kilowatt of electric power output. Panasonic is emphasizing the relatively long life of its units, suggesting that its products will work well for up to ten years after installation. But this is still a shorter life than a conventional home heating boiler that burns gas to heat water, so consumer acceptance is not guaranteed.

WHAT FUEL CELLS CAN DO FOR EMISSIONS

Emissions from the use of fossil fuel energy to heat, cool, and power buildings are as much as half of the total greenhouse gas output in most developed countries. Fuel cells offer huge potential for slashing these emissions.

A solid-oxide fuel cell powered by natural gas may reduce the greenhouse gas output of a home by 30 percent or more, a much larger reduction than seen in early domestic combined heat and power units using internal combustion systems. But even the best performing fuel cell would still leave the typical northern European house burning enough natural gas, mostly for winter heating, to produce 4 tons of carbon dioxide emissions. So why are solid-oxide fuel cells such an important potential advance in the move toward a low-carbon future?

The answer is that eventually we will use renewable fuels to run these cells, rather than natural gas. In fact, one of the many advantages of solid-oxide cells is that they can be fueled by a whole range of hydrocarbons, including cellulosic ethanol, the second-generation biofuel discussed in Chapter 7. If we power fuel cells this way, we will reduce net emissions to a very low level indeed, perhaps as little as 10 to 15 percent of the impact of natural gas. Fuel cells will also be able to exploit fuels from other renewable sources such as methane from slurry heaps or from sewage treatment plants.

A further advantage of fuel cells is that the waste products will usually only be carbon dioxide and water vapor. The vapor can be condensed back into a liquid pure enough to drink. The carbon dioxide then forms 100 percent of the waste stream and therefore can be relatively easily sequestered. (As Chapter 8 shows, one of the most difficult tasks in a carbon capture and storage process is separating the stream of waste gases to create almost pure carbon dioxide.)

This raises a very interesting possibility. At some stage, we should be able to use fuel cells to construct a fuel cycle that is "carbon-negative"—that is, better than carbon neutral. The cell will use ethanol made from renewable energy crops, such as switchgrass or wood wastes, which have absorbed carbon dioxide from the atmosphere. If the carbon dioxide produced during the operation of the fuel cell can then be captured and permanently sequestered, the whole cycle could actually result in a net removal of greenhouse gases from the atmosphere. This beneficial outcome is not something that will happen in less than a decade. Difficult technical issues must be resolved, such as how to safely and cheaply compress and liquefy carbon dioxide on a relatively small scale. Nevertheless, this method may eventually become one of the cheapest ways of reducing existing atmospheric carbon dioxide levels.

Fuel cells have another environmental benefit. Unlike fossil fuel generating plants, they produce almost no emissions of other polluting gases such as oxides of nitrogen or sulfur compounds. (A solid-oxide fuel cell powered by natural gas will scrub away the small amount of sulfur before the fuel is separated into carbon monoxide and hydrogen.) In countries with unacceptable urban air quality, usually partly caused by old or inefficient coal- or oil-burning power stations, the fuel cell offers the prospect of alleviating dangerous atmospheric pollution.

Nevertheless, we should not overestimate the attractiveness of replacing gas boilers with fuel cells in domestic homes. For one thing, the issue of space is unresolved. A cell with an integrated top-up boiler requires lots of room in itself. If it were to be powered by ethanol, the homeowner would also need space for a storage tank. Per unit of energy, ethanol takes up about 50 percent more space than fuel oil, so the tank would need to be very large or be frequently replenished.

These are important obstacles to fuel cells becoming the dominant source of domestic heat and power, but large-scale cells for offices and factories do not suffer from the same disadvantages. Take data centers, for example. These buildings, containing racks of computer servers connected to the Internet, now use over 2 percent of the electricity produced in the world. They're putting a strain on electricity supply, and in some places, such as parts of California, they've even had their power supply capped. Data centers need power both to run the thousands of servers in the building and to provide cooling to prevent the computers from overheating. They have high electricity demands twenty-four hours a day and (with the exception of a few newer centers with fresh-air cooling) they need substantial amounts of energy for cooling almost all of the year.

A large fuel cell power plant attached to a data center would be the perfect solution. The cell could provide the electricity for the servers, while the heat created could power adsorption chillers to cool the building. This approach would prevent virtually all the waste associated with centralized electricity generation and provide the data center with secure and reliable power. Of course, the building would still be connected to the wider electricity grid so that in the event of the fuel cell failing, power would always be available.

Moreover, although homeowners might find having a large ethanol tank inconvenient, operators of large buildings will have no such problems. Indeed, they may well already store diesel fuel to provide backup power to protect against power cuts and other interruptions in supply. So keeping a stock of liquid fuel will involve very few extra costs. For major electricity users such as data centers, supermarkets, hospitals, and large high schools, the benefits of using fuel cells powered by renewable liquid hydrocarbons such as ethanol are overwhelming. Probably installed and maintained by the local power utility rather than the building owner, they promise to provide reliable, genuinely low-carbon power at reasonable prices.

Generating companies will also see big advantages to having large commercial buildings powered by onsite fuel cells. Not only will it enable them to serve extra customers in areas of tight supply, but it will also give them a substantial source of replacement power at times when renewable electricity supply into the grid is limited. A large fuel cell in a school, for example, would face only minimal onsite demand in the evening. So it could either throttle back production or work flat out and export the excess power to the local grid. The utility company could have full remote control over the cell and increase electricity output at times of general power shortage. To illustrate precisely this point, Ceramic Fuel Cells has already successfully demonstrated that it can remotely adjust the output of one of its micro units from a control room tens of thousands of miles away. A small number of electricity utilities have begun to conduct trials to test how fuel cells can be automatically used to adjust their power output at times of peak demand. If domestic and industrial cells can be turned up to full power when the power companies are short of electricity, they might represent another important buffer that allows the grids to accommodate larger and larger amounts of intermittent power from the wind or sun.

Solid-oxide fuel cells do not respond immediately to a call for more power: they can take ten or twenty minutes to adjust the electricity output to what is required. But this characteristic makes them perfect for matching the highly predictable daily swings in tidal power or the likely variations in solar energy. Concentrated solar power will give us daytime electricity. As the power of the sun ebbs away at the end of the day, fuel cells can be gradually ramped up to full output. In combination with other technologies that can provide power almost instantaneously, such as the pumped hydroelectric storage described in Chapter 1, large fuel cells offer a really substantial insurance as national electricity infrastructures become more exposed to the variations of renewable generation.

All told, then, fuel cells have enormous promise. They're still not competitive with large coal-fired power stations in terms of cost per unit of electricity generated, but we can be reasonably sure that much of our power, heating, and even cooling will be eventually generated in fuel cell plants attached to homes, apartment blocks, and commercial buildings. In time, these cells will be powered by renewable fuels, such as cellulose-based ethanol, making them even more environmentally attractive, and maybe even carbon-negative.

DISTRICT HEAT AND POWER

Fuel cells are scaling up to power and heat large buildings, but the other approach to combined heat and power seeks to supply entire urban areas. The story begins with district heating plants, which developed not because of fears over climate change but because they offered households a relatively cheap way of obtaining heat in winter. Particularly in towns far from major gas networks, a centrally located heating plant, often powered by local wood, provides a secure and inexpensive means of keeping the population warm. Hot water is distributed to homes in insulated pipes fanning out from the heating plant. Users can adjust the flow of hot water through their radiators in the same way as they can change the settings in conventional central heating systems. The more hot water they use, the more they pay.

The prices charged vary from town to town, but most suppliers charge the equivalent of 5 to 7 cents per kilowatt-hour of heat. If you factor in the savings made by not having to install and maintain an expensive central heating system, this works out to be broadly competitive with traditional gas heating and cheaper than using oil. So most district heating systems around Europe offer homeowners reasonable value for money. Unlike solid-oxide fuel cells, they're already competitive with fossil fuels.

Increasingly, district heating plants are being used to generate electricity as well as heat. Like fuel cells, an efficient district plant can convert almost all the energy in a fuel into a mixture of power and heat. Little energy is wasted, and energy losses from carrying the hot water or electricity to homes are limited. Best of all, if the plant burns renewable fuels, such as local wood or municipal waste, it has only minimal greenhouse gas emissions.

Fuel cells and district heating plants have much in common, but they will be used in very different ways. Fuel cells are exciting because, thanks to recent advances, they can convert over half the energy value of gas into electric power. They can be put in a house or hospital, and the secondary output, heat, is essentially

123

free. If the heat (or cooling power) is used, then all is well and good. If it is not, then little is lost. Ideally, the fuel cell runs all the time, sending electricity to the wider grid when its output is not required onsite.

District heating plants are different. They are not very efficient at converting fuels into electricity, perhaps providing only 20 percent of their energy output in the form of power, with the other 80 percent being heat. This means district heat and power plants are usually operational only when they are able to sell heat—or the equivalent amount of cooling power. Many future plants should get around this problem thanks to gasification technology. First, the fuel is heated to about 700°C (1,290°F) in a low-oxygen environment, just like the charcoal-making process discussed in Chapter 9. The heat drives off hydrogen and carbon monoxide, both of which can be used as fuels to drive turbines or heat water. Once this process is complete, the remaining charcoal can be burnt. One of the first large-scale biomass gasification plants in Europe has been successfully operating in the Austrian forest town of Güssing for several years, producing 2 megawatts of electricity and 4.5 megawatts of heat for the district heating system. The ratio of electricity output to heat may make running the plant economical even when the heat is not needed.

Some countries use district heating extensively, others barely at all. The U.K. and the U.S., for example, rely largely on central heating boilers in each home. Canada has about 150 working district heating schemes, mostly serving city center buildings. However, Charlottetown, Prince Edward Island, started its own local hot water system in 1986, burning wood chips from otherwise unusable local timber. The system also generates electricity using a steam turbine. The network provides heat to eighty town buildings, including university, hospital, and government offices as well as apartment buildings and a small number of private homes. The heating systems in North America are generally small in extent and cover few homes.

In Denmark, by contrast, almost two-thirds of the population gets its hot water and heat from over four hundred district heating systems. Some of these are cooperatively owned ventures, serving a few hundred households, while the biggest provide heat to 100,000 homes or more. About three-quarters of the plants in Denmark also generate electricity. The plant burns fuel and generates steam, and some of this steam drives electric turbines, while most of the hot water is separated off for heating. The district heating trade association claims that more than 90 percent of the energy in fuel is typically converted to usable heat or electricity. Most Danish plants are still fueled by gas or other fossil fuels, but about 40 percent of the heat produced doesn't result in a net increase of carbon dioxide, because the plant has burnt renewable wood or domestic waste. (If the waste had been buried instead of burnt, it would have decomposed in a landfill site, producing methane and other greenhouse gases, some of which would have escaped to the atmosphere. Since methane is a worse greenhouse gas than carbon dioxide, there are clear climate change benefits to using waste food in this way.)

The carbon savings from using district heating in Denmark are said to be substantial. An independent trust that works to improve energy efficiency says that the typical Danish district heating scheme produces heat with only a quarter of the carbon footprint of heating a home by electricity. A homeowner moving from electricity to district heating will generally consume more heat because it is cheaper. But the increase in consumption is typically only about 15 percent, so it wipes out only a small fraction of the carbon savings.

On the other side of the Atlantic, Canadian company Nexterra is pushing gasification technology for large-scale plants designed to burn forest wastes. According to one estimate, almost half of all material taken out of forests is not used to make timber or pulp and is therefore available as fuel for decentralized power stations. One project sees Nexterra's gasification technology, praised by industry insiders for being simple and reliable, installed as the heating plant for a new urban community in the port area of Victoria, British Columbia. This innovative technology helped win the new community an award from the Clinton Climate Initiative in July 2009 as one of just sixteen Climate Positive Developments worldwide, recognizing that the Nexterra plant provides a net surplus of energy for export to the wider area. The plant is flexible enough to provide just heat, just electricity, or a mixture of the two. Nexterra is also going to provide a biomass gasification plant for heating the campus of the U.S. government's premier energy research center at the Oak Ridge National Laboratory in Tennessee. Another proposal takes gasification technology to fifteen communities in Canada's western interior, where the wood stock has been seriously affected by pine beetles that kill the trees and reduce the value of the timber. The network of wood gasifiers would each generate up to 10 megawatts, providing power for thousands of homes in communities whose economic viability has been seriously affected by the destruction caused by the beetles.

How big a difference could wood-powered CHP plants make to our low-carbon future? The potential is huge, but the availability of land to grow trees remains a key issue. Take Sweden, a country where many municipalities use district heat and power plants. For example, the plant in Borås, near Gothenburg, provides about 25,000 homes and 2,000 offices with most of their heating needs and much of their electricity. To do this, it uses about 270,000 tons of wood chips a year, or approximately 10 tons per customer. Sweden as a nation produces about 30 million tons a year of wood from its forests, or only about 6 tons per household. Therefore even if *all* the Swedish wood harvest were used for district heat and electricity, there wouldn't be enough to keep everyone in the country warm. If one of the world's major exporters of forest products does not produce enough energy in its timber, surely very few other countries will have enough raw material either.

This interpretation is too harsh. In truth, the amount of wood taken from Sweden's forests is governed by world demand for paper and timber, not by a shortage of woodland. Sweden boasts several billion tons of standing trees, compared with an annual need of less than 50 million tons to keep its entire population warm with district heating. Indeed, the European continent as a whole has almost 4 million square miles of forest land, and if 20 percent of this land were used to grow fast-growing species of trees, such as willows in damp temperate countries, it would provide enough energy to heat all European homes at today's patterns of consumption. In a densely populated country like the U.K., with a relatively small area of forest, about 10 percent of the entire land area would need to be given over to growing wood for fuel, though this figure could be reduced substantially by improving the insulation of existing buildings, as discussed in Chapter 5. In the case of Canada, the numbers are even more compelling. About a third of the country's land mass—about 740 million acres—is thickly forested, and about half this land is accessible. Less than 2.4 million acres are harvested each year. Sustainable use of northern forests could meet all of Canada's heat needs and a large fraction of the United States'.

However, the problem is that heating plants aren't the only source of new demand for wood. Two other chapters of this book—those on cellulosic ethanol and biochar—also focus on prospective uses for woody matter grown on the world's limited supply of reasonably fertile and well-watered land. The crucial question is how best to balance the primary requirement of using land for growing food for an increasing population with using land to grow energy crops. The striking food price increases of 2007–2008 provided a sharp reminder of the inherent conflicts between food production and the world's need for energy, as corn crops were diverted to ethanol production for cars.

In Chapter 10, I propose that one answer to this conflict may be to renew our focus on improving the huge areas of the world with degraded soils on which virtually nothing grows. Any real solution to climate change must involve restoring soil health and rolling back desertification, probably using massive tree-planting schemes such as those used in western China. Increasing the amount of usable land on the earth's surface will help us meet the challenge of growing more woody biomass and feeding 3 billion more people than the globe sustains at present.

5

SUPER-EFFICIENT HOMES

Passivhaus and eco-renovations

IN 1991, a terrace of four new homes was completed in Darmstadt, Germany. From the outside, they looked much like similar houses completed in Germany at the time, but there was a
crucial difference. None of the homes featured a conventional
central heating system. This wasn't an oversight or an exercise
in promoting ascetic living. These houses, designed by a group of
Swedish academic architects led by a soft-spoken German engineer called Wolfgang Feist, were the first "Passivhaus" buildings,
which means they were so well insulated and cleverly designed
that they didn't need a full central heating system, nor, indeed, an
air conditioner for the summer.

Now, almost twenty years later, the philosophy behind the
Passivhaus is spreading around the world. North America only
has a handful of true Passive Houses that meet the rigorous technical standards set by the European originators, but interest is
growing quickly. Root Design Build in Portland, Oregon, began in
summer 2009 to construct a Passive House that aims to exceed
the tough energy-efficiency requirements. It will be the first one

on the U.S. West Coast, says the company. But over ten Passive House homes are now in development throughout Oregon, and the U.S. arm of the German Passivhaus Institute had trained over 150 architects by the end of 2009. The new house, named the Shift House to convey something about the radical shifts in construction techniques and materials used in the building, is not going to be cheap compared with similarly sized standard homes. At $330,000 for a 1,700-square-foot building, the financial arguments for this expensive prototype depend on reducing energy bills. The aim is to cut heating and electricity costs to no more than about 10 percent of the average American house of the same size. That means, said one of the designers to the local newspaper, a target heating bill of just $194 a year.

Despite this eye-catchingly low figure, heating and cooling don't generally get quite the same media attention in the climate-change debate as cars and electricity-hungry gadgets. But they should. If you added up the emissions of all the world's gas and oil boilers, coal fires, electric heaters, and air-conditioning units, then you'd probably find that managing the temperature of buildings—either through heating or air conditioning—is the world's single most climate-damaging activity.

A large slice of heating and cooling emissions is created needlessly, since almost all homes and offices are a very long way from Passivhaus levels of insulation and intelligent thermal design. Much of the hot and cold air created by boilers and air-conditioning units is hemorrhaged through leaky walls, windows, floors, and roofs. This is a huge economic waste as well as an environmental problem, with soaring energy costs making it more and more expensive for homeowners and businesses to maintain a comfortable temperature.

Over the last twenty years, most developed countries have introduced regulations that demand increasingly high insulation standards when new buildings are constructed. But governments around the world have been surprisingly slow to push for better energy efficiency in existing buildings. This is a mistake: in the U.S., the number of new homes constructed each year may be less than 2 percent of the number of houses in the existing stock. The need to get homeowners to refurbish their existing homes is therefore far more urgent than imposing higher and higher insulation standards on new buildings. Unfortunately, imposing regulations on construction companies is easier and less intrusive than instituting a massive eco-renovation program. With a few exceptions, governments are avoiding dealing with the obvious need to substantially improve older buildings.

Nevertheless, there is reason for optimism. Unlike some of the technological innovations in this book, many domestic energy-efficiency measures make financial sense even in the short term. We don't need further technological improvements or a high tax on carbon emissions. For example, at today's energy prices, it is often sensible for the homeowner to very significantly improve the insulation standards of the home, especially in the countries where winter temperatures are very low. International experience, particularly in Germany, is that a wide-ranging program of education, encouragement, subsidy, and cheap loans can successfully push landlords and homeowners into taking action. Implemented enthusiastically, this program can yield energy savings of over 60 percent, perhaps as high as 80 percent. One important eco-renovation in Austria of a 1950s apartment block cut energy use by 90 percent.

There really isn't any alternative to improving the efficiency of existing houses. We can't simply tear down the hundreds of millions of leaky homes that exist today—we need to find ways of reducing energy use while leaving the fabric of the building intact. Doing so will raise aesthetic issues: the most effective way of reducing energy use is to introduce a thick layer of insulation on the exterior of the walls of houses, which may change the building's appearance and incite strong local opposition. Insulation invisibly attached to the inside of the house can be nearly

131

as effective, but it does reduce the size of the rooms. Combined with better windows, floor and loft insulation, and good central heating boilers, eco-refurbishments could cut carbon dioxide emissions by a substantial percentage.

Before looking at renovations, however, this chapter explores the ideas behind the Passivhaus movement. Although only a few thousand houses have been built to Passivhaus standards, these buildings have shown that huge reductions in energy use can be designed into all houses, not just expensive eco-homes. Good construction techniques and ruthless attention to detail matter as much as the choice of insulating material. Passivhaus thinking has become embedded in new building activity in several different parts of the world, but it can also inform the massive program of eco-renovation that the developed world urgently needs.

PASSIVHAUS

The Passivhaus idea is simple. A house insulated to the highest standards does not actually need a central heating system. Even in the depths of winter, it can be kept warm by capturing energy from the sun and from the heat given off by the people and electrical appliances it contains. On the coldest days at high latitudes, the building may need a top-up from an electric radiator, but even in cold countries, a well-built house can remain comfortably warm during winter. In hot climates, Passivhaus construction can help dramatically reduce the need for electric air conditioning.

The first houses to meet the Passivhaus ideal were constructed in Germany almost twenty years ago, so the idea has taken a long time to blossom. Even now, there are probably fewer than fifteen thousand certified Passivhaus homes around the world, and most of them are in Germany and Austria. The slow adoption is surprising. A properly built Passivhaus dwelling should use less energy for heating than 10 kilowatt-hours a year for each square meter, or a little less than a kilowatt-hour per square foot, of floor area. The Passive House that Root Design Build is constructing in Oregon should consume less energy, in total, than 1.4 kilowatt-hours per square foot, compared with a U.S. average of around 12 kilowatt-hours. Getting from the levels of energy use typically seen around the world down to the levels that can be achieved by full eco-refurbishment will save several tons of carbon dioxide per house, or at least as much as completely eliminating the greenhouse gas emissions from the household's car.

The intellectual force behind the Passivhaus standard remains the German engineer who built the first Darmstadt house, Wolfgang Feist. He went on to found the Passivhaus Institute, a body dedicated to setting the standards for energy efficiency in home construction. Feist is an engineer, not primarily an architect, by training. Indeed, listening to one of his talks, you quickly understand that reducing the carbon emissions from housing is largely an engineering challenge and has relatively little to do with architecture, at least as it is conventionally understood. Passivhaus homes do not need to look any different from the prevailing architecture of the area, and they certainly don't have to be small or strangely shaped. How a house looks isn't important: to get Passivhaus certification is simply a matter of meeting energy use requirements. Neither does a Passivhaus have to be high-tech, full of steel, concrete, and granite and controlled by sophisticated electronics. And although a Passivhaus will often use solar collectors to heat water, there is no need for expensive photovoltaic panels or domestic wind turbines.

According to Wolfgang Feist, achieving energy efficiency in new housing simply requires the builder or renovator to focus on five key principles: excellent wall insulation; small, high-quality windows; airtightness; a lack of "bridges" that conduct cold into the house from the outside air; and a ventilation system that brings fresh air into the house and preheats it using warm, stale air extracted from the main rooms.

Putting all of these elements together is not a simple matter and can be expensive if done without thought. But there is no magic or unusual technology involved. Let's look at each principle in turn.

Wall insulation

Walls are the main source of heat loss in most homes. Although huge amounts of energy can be lost through the roof, most houses have sufficient loft insulation to reduce the outflow of heat (though almost all would benefit from another layer). Walls are a more difficult problem. To meet Passivhaus standards, a home constructed from bricks will need a thick layer of insulation either on the exterior wall, in which case the facing brick will be invisible, or on the inside walls of the house. This insulation will need to be around 16 inches thick, significantly reducing the internal dimensions of a room, if installed internally, or adding to the bulk of the house if used on the exterior walls. This insulation will usually be made from expanded plastics, but a wide variety of alternatives are available.

Windows

Even very well-insulated windows let in more cold than a wall, so an energy-efficient house needs to have a relatively small percentage of its surface area given over to glass. This restriction doesn't need to make the house dark; some of the brightest homes I have ever seen have just been constructed on a large eco-estate in Milton Keynes, north of London. Although they weren't built to Passivhaus standards, they are better insulated than any other mass-produced U.K. houses. The windows were intelligently positioned to capture as much light as possible, particularly in winter.

The Passivhaus approach is to face all large windows to the south (or to the north in the southern hemisphere), which maximizes not only the incoming light but also the heat from the sun caught by the house in the winter months. The Passivhaus standard looks for 40 percent of the total winter heating need to be met from the sun's heat entering the house through window glass. If a shade is put above the south-facing windows, the high angle of the summer sun means that relatively little unwelcome heat is captured in the hottest months of the year. When the sun is high in the sky, it is most important to have a well-insulated loft that blocks the heat from entering the house through the roof.

Wolfgang Feist points out that perhaps only 70 percent of a window is glass. The rest is the frame and the fittings. Energy losses from these elements can be far worse than from the glass, so considerable effort has gone into designing frames that do not leak heat. This task sounds simple, but the engineering is actually very complex. Such windows are usually triple glazed, with inert gas inside and glass surfaces that reflect heat back into the house. Even these windows probably emit five or six times as much heat as a really well-insulated wall, so they can't be too large.

Feist gives some illuminating figures for the impact of good window design on levels of internal comfort in the winter. A well-made, triple-glazed, argon-filled window will feel warm even when temperatures are well below freezing outside. He says that the best examples can keep the temperature at 18°c (64°f) on the inside of the glass, compared with just 5°c (41°f) for a traditional double-glazed window.

The difference this makes to the feeling of comfort in the room is very marked. A room with an air temperature of 19°c (66°f) but with warm windows will often feel more comfortable than a room at 21°c (70°f) but with cold temperatures at the window glass. The reason is that the warm human body loses radiant heat to the cold window. Further, if a person stands or sits sideways to the window, he or she will give up more radiant energy on one side of the body than the other, which tends to lead to even more discomfort.

135

"Bridges" that conduct cold into the house
Conventional construction techniques often allow very conductive materials, such as metals and concrete, to provide a bridge

between the cold outside air and the inside of a house. Even the best-built homes often lose significant amounts of heat this way because of poor design and carelessness during construction. Passivhaus homes avoid the problem by using carefully pre-fabricated components and vigilance during the building process. Factory prefabrication of houses—still unusual in some parts of the world—helps reduce heat losses because components are made to more accurate specifications in clean and dry conditions in a factory. Passivhaus homes don't necessarily have to be factory made and then assembled on the building plot, but this is the easiest way to achieve the standards required.

Airtightness

Even the best conventional homes lose huge amounts of warm air through cracks, poor door seals, and other routes. New construction techniques vary around the world, but few builders anywhere understand how important airtightness is to the overall energy consumption of the home. For example, one U.K. government body has said that today's "best practice" still produces over three times the level of air loss allowed by the Passivhaus standard. In a typical new house, perhaps 30 or 40 percent of the heating requirement arises because of the ingress of cold air through gaps created accidentally during construction. To get a Passivhaus certificate, the building must pass a test in which the air pressure in the house is increased and the rate of air loss to the outside world is measured. To pass the test, the house must lose less than 60 percent of the volume of air in the house per hour. By comparison, an older conventionally built house will often have ten or twenty times this rate of leakage, particularly in windy locations. Those who cannot believe this figure should sprinkle a fine powder, such as talc, at the corners of an older room on a windy day and watch the moving air blowing it around.

Creating a fairly airtight new home is difficult, even if the components are factory made and fit together very tightly. The home-building industry doesn't tend to attract finicky per-fectionists used to working to hundredths of an inch, but that attention to detail is what is required to get a Passivhaus building to achieve its full potential.

The ventilation system

Because a Passivhaus home is so airtight, it needs to have a ven-tilation system that brings fresh air into the house and extracts the stale air. Otherwise, the occupants would suffer from excess carbon dioxide, which builds up as a result of human breathing. The house would also suffer from pollution from other sources such as the unpleasant chemicals given off by most paints. But a simple air-extraction system using fans would be no better than a leaky house: warm air would leave, and cold air would come in.

The Passivhaus solution is the use of heat exchangers: cold air entering the house passes over ducts containing the warm (and humid) air leaving the building. This kind of ingenious heat recovery system can transfer a remarkable 80 percent of the out-going heat to the incoming air and also provides the ideal place to top up the heating, when required. Electric elements can heat the incoming air on the few days a year that a bit of extra warmth might be necessary. Typically the air in the whole house will be changed every couple of hours, ensuring abundant ventilation as well as excellent energy efficiency.

Passivhaus principles also work in hot countries. A thick and effective barrier of insulation combined with airtightness and forced ventilation can all work to keep the heat out of a house just as well as they ensure winter comfort in cold countries. Although few certified Passivhaus homes have so far been built in the tropics, the Passivhaus principles make perfect sense in such regions. The key difference between northern Germany and Australia might be the way that air is cooled before coming into the house. In a very hot country, it could, for example, be passed along a duct running 6 feet below ground. In the middle of the day, temperatures well below the soil surface are much lower than those in the air.

The pioneering Passivhaus homes in Darmstadt have been the subject of much research during the past two decades. One key finding is not only do residents have extremely low heat demand, but electricity use for appliances and lights is also well below German averages. In one sense this is not surprising: we might expect people who live in Passivhaus homes to be interested in energy efficiency. And indeed, some researchers have suggested that a large part of the total energy savings arise because homeowners are highly motivated to run the house efficiently. Wolfgang Feist points out that the evidence tends to contradict this hypothesis: there is as much variation in energy use between different Passivhaus homes as there is between conventional houses. Some Passivhaus homeowners are relatively profligate in their use of energy, and others are extremely careful. So the low average energy use does not result simply from the occupants neurotically trying to reduce their energy bills. The houses are bright, so the need for electricity for lighting is low. The absence of a central heating system also reduces power use because there are no pumps and controls. Passivhaus standards really do radically diminish the average use of energy in all types of houses and in all temperature zones.

It is difficult to come up with many substantial disadvantages to the Passivhaus approach. Perhaps surprisingly, for many people, the most troubling effect of living in a Passivhaus home is that most external noise disappears. If the windows are closed, the sounds of birdsong, light traffic, or children playing in the street are all absent. Another difference is that the air in the house tends to feel very dry. Air in homes is made fairly humid by human respiration and by water vapor from kettles, showers, and cooking. In a Passivhaus home, the mechanical ventilation system extracts this wet air and replaces it with colder air from the outside. Cold air can hold very little water vapor compared with hotter air, and external air coming into the house at −5°C (23°F) will often be nearly saturated, containing as much water as it can possibly hold. The humidity level of this air falls when it is heated by the heat exchanger in the ventilation system. Air at 20°C (68°F) can hold five times as much water vapor as air at −5°C, so the incoming air is now only 20 percent saturated with water vapor. Human beings are more sensitive to this relative humidity level (how much water vapor the air contains compared with the maximum it could contain at that temperature) than to the actual amount of water in the air. Most people feel comfortable with relative humidity levels over 40 percent, so the Passivhaus mechanical ventilation system will inevitably make the house feel dry when the temperature is very cold outside. This effect is exacerbated if the air is dusty. So the owners of Passivhaus homes need to invest in good vacuum cleaners as well as humidifiers and large numbers of potted plants to maintain moisture levels at all times. Neither of these two disadvantages seems particularly off-putting.

The need for substantial extra blocks of internal insulation means that the rooms are very slightly smaller than they would otherwise be, but the effect is marginal. In one house I looked at, the thicker walls had reduced the internal space by less than 3 percent. That's barely noticeable in a large house, and it's partly offset by the space saved by not having radiators on many of the internal walls.

How much do Passivhaus homes cost, and what are they like to live in? I spoke to a small housebuilder in the remote west of Ireland to get a view from a country that has only recently started to build them. Scandinavian Homes imports extremely well-insulated prefabricated housing from Sweden and decided four years ago to offer a Passivhaus upgrade to its existing product line. It built its first demonstration house in 2005 and has been constructing Passivhaus homes for sale since 2006. By a fortunate chance, the person who answered the telephone when I called the office happened to be the company's first Passivhaus customer. After her home was constructed, she eventually became an employee of the business. Miriam Green's family moved into their Galway home on the windy and wet western coast of Ireland in 2006. "The first winter wasn't easy," she said. "We thought we'd get enough warmth from the electric appliances and from body heat, but we were wrong. We didn't realize that the rules allowed us to use some under-floor electric heating in the depths of winter as well. But our usage of heat generated by electricity never went above the Passivhaus standard of 10 watts per square meter." (This is less than 10 percent of a typical Irish home.)

Green's house is large—almost double the average size of a European home—so it is not surprising that heat from the occupants and the electric appliances is not quite enough to keep the building warm on the coldest days. Another issue for houses in cloudy western Ireland is that the winter climate offers little solar warmth. Although temperatures don't often go much lower than freezing, the overcast days reduce the amount of solar energy coming in through the south-facing windows. The 2007–2008 winter was particularly cloudy, and Green noticed the effect on the temperatures in the house. She also commented on another problem: the roof windows of her home, imported from a large manufacturer in Scandinavia, are simply not well enough designed. They are triple glazed, but the seals around the edge do not fit snugly, and on the windiest days she can feel a draft (a problem that chimes with Wolfgang Feist's passionate focus on improving window frames). Green says, however, that the single-story Passivhaus homes that her company offers never suffer from this problem because they don't use roof windows.

The amount of electricity that Green's family uses reflects the high insulation standards. She and I easily calculated that, in the coldest month of the past year, the house must have used about 800 kilowatt-hours, of which well over half had probably been used to run the home appliances. Heating demand was perhaps a tenth of a poorly insulated house of the same size, very much in line with Passivhaus expectations. But electricity is a very expensive way of heating a house—perhaps four times the price of gas for an equivalent amount of energy—so the savings in cash for Green's family are not likely to be as great. And as an extremely ecologically aware individual, Green was concerned that heating a house with electricity, albeit in relatively small amounts, was bad practice because of the carbon dioxide implications of power generation. The figure varies between countries depending on the type of fuel used to generate electricity, but a kilowatt-hour of electricity usually produces two or three times as much carbon dioxide as a good gas boiler delivering the same amount of energy. So she is looking to install an innovative heat pump to replace the electric under-floor heating. Heat pumps still need electricity for their power, but a modern system will produce three units of warmth for every unit of electricity that they that they use.

The homes built by this small Irish company were already highly efficient and were engineered to be airtight even before they offered a Passivhaus upgrade. The mainstream houses of Scandinavian Homes aren't representative of the average cost of constructing a house in Ireland, so it is not easy to work out the extra cost of building to Passivhaus standards. Nevertheless, in her role as employee of the firm, Green gave me some estimates. She said that she thought the materials cost of a typical Passivhaus bungalow was about 5 percent higher than a conventional home from her company, but a two-story house might cost 15 to 20 percent more. Most of this incremental cost arises because of the extra internal insulation. However, not needing a central heating boiler and room radiators means a big savings. All told, Miriam reckons that upgrading a very well-insulated house into the Passivhaus category probably adds in the region of 10 percent to the aggregate cost of materials and labor.

These figures are consistent with the estimates provided by Wolfgang Feist of the Passivhaus Institute. Based on German prices, Feist says that the total cost of building a Passivhaus is about $22,000 greater than for a comparable home constructed to today's government-mandated insulation standards. Subtract the savings made by not having a central heating system, and the incremental cost works out to about 8 percent of the average construction costs of a new German house. Not only will such a house provide comfort during the winters and hot summers, but the lower energy bills would probably pay back the investment in a dozen years or so. In other words, anybody wanting a new house would be very well advised to buy one built to Passivhaus standards.

So why aren't more houses being built the Passivhaus way? Miriam Green answers from her own experience. Her family found the prospect of a Passivhaus home intimidating. "It required a huge leap of faith," she said. "We were the first people in Ireland to commit to buying a Passivhaus home, and we took some convincing that the house would be warm." But now, as an employee of a pioneering building company, she finds attitudes are changing fast. "People in Ireland now understand what a Passivhaus is," she says. "They're much easier to sell now."

There's also another problem. The housebuilding industry simply isn't used to constructing energy-efficient homes. To start building to Passivhaus standards will completely change the way large housebuilders do business. Until the workers on a construction site have been fully trained, every single phase of the complicated process of building a house will have to be closely supervised. Many of the construction techniques that the industry has evolved over the last half century to build homes quickly and inexpensively with relatively unskilled labor are simply not compatible with the relentless attention to detail required by the Passivhaus approach. I talked to two large housebuilding firms in the U.K. that had pioneered small developments built to high-insulation standards, though neither had yet constructed any housing to full Passivhaus rules. Both told me that their first homes had cost well over 40 percent more than comparable developments elsewhere. They acknowledged the cost increase arose because of inexperience, mistakes in design, and problems in sourcing the unusual materials. But there is a learning curve in housebuilding just as much as in the manufacturing of wind turbines, and as housebuilders get more experienced, costs will fall to the level of those of the small builders who have pioneered Passivhaus construction.

ZERO-CARBON HOMES AND ECO-RENOVATIONS

Interest in super-efficient homes is there, but low-carbon housing is really not taking off at the pace that it needs to if we are to see a substantial reduction in overall emissions from the housing stock. Even in Germany, Passivhaus construction accounts for only a small proportion of total building, although some people talk about getting it up to 20 percent of all new buildings within a few years. The lack of progress means that countries around the world are now using the law to force builders to start constructing really energy-efficient housing, even though customers ought to want them anyway because of the lower energy bills.

The U.K. government has set the astonishingly ambitious target, one unmatched by any other country, that all new housing in the U.K. must be "zero-carbon" or "net-zero"—meaning that any energy derived from fossil fuels used in the house must be balanced by renewable energy generated on the same site or very nearby—by 2016. So whereas the Passivhaus principle reduces heating use to perhaps 10 percent of that in a conventional house and may cut the electricity needed to power the lights and appliances in half, the U.K. rules for new construction will become far more demanding. The government has set a huge challenge but one that is theoretically possible to achieve.

British construction companies have divided into those that think the new rules will make new housing impossibly expensive

143

and those that see the targets as an interesting way of stimulating a rather conservative industry into rapid change. One of those more optimistic businesses is the Irish building materials firm, Kingspan. Its U.K. subsidiary has produced plans for a "self-build" house—constructed by tradespeople that the purchaser of the new building manages—that is zero-carbon because it has many solar photovoltaic panels on the roof combined with very high insulation levels and good airtightness levels. A wood-burning boiler provides supplementary heat, and because wood is classed as a renewable fuel, the carbon dioxide output is not counted in the zero-carbon calculation. However, this house, called the Lighthouse, is extremely expensive. It costs about $300 per square foot to build, well over twice the construction cost of a less energy-efficient building on the same site. For comparison, some Passivhaus buildings in Germany have been built for less than $150 per square foot, a figure that is much closer to the lower construction costs generally seen in the U.S. The Oregon Passive House described in the first pages of this chapter will be about $200 a square foot, far higher than the local average.

From the outside, the Lighthouse is extraordinarily attractive, looking like the spinnaker of an oceangoing yacht in full sail. Inside it seems a bit cramped, particularly in the bedroom areas on the bottom floor of the house. Light levels around the building are good, despite the relatively small percentage of the wall area given over to windows. But I would strongly doubt that the typical new home buyer in the U.K. is going to willingly spend an extra $160,000 to live in an ultra-low-carbon house.

144

Moreover, when it comes to carbon emissions, the simple fact is that this $160,000 could be much better spent. The law of diminishing returns applies. A net-zero house will likely always be far more expensive than a house built to the Passivhaus standard, but the incremental savings in carbon emissions are low. Using the money to renovate the oldest buildings and take them up to modern standards of insulation and airtightness would have many times as much impact.

I talked to Tim Fenn, who runs a building firm that focuses on energy-efficient renovation near Oxford. He said that a typical large and drafty Victorian house in Britain, owned by a family wanting to maintain the traditional external appearance of bricks and mortar, might consider two different types of improvement. Taken together, these improvements might cost $40,000 and save 75 percent of the heating bill in an old house—and around 4 or 5 tons of carbon dioxide a year. If the house is being fully renovated, the first improvement would be to install sheets of insulation on the inside of the exterior walls. And, to improve airtightness and reduce heat loss, the renovator can install a thin insulation material faced with aluminum foil to block the flow of radiant heat and air from the room.

Heat moves from a hot place to a cold place by three mechanisms: radiation, convection, and conduction. A layer of half-inch-deep, foil-faced insulation does relatively little to stop heat conduction but is very effective at blocking infrared radiation. Experts disagree on just how much of a house's heat loss is accounted for by radiation, but some claim it is as much as one-half. Just adding reflective foil membranes to the inside of the main rooms in an eco-renovation might prevent a third to a half of the heat loss from an old house. It would also keep the house cooler in summer because the aluminum would help block solar radiation from entering the house.

The poorly insulated North American housing stock would benefit from substantial renovations of the type that Tim Fenn identifies. In places where old, historic buildings are a vital part of the landscape, homeowners may be reluctant to add insulation to the exterior of their houses. Any added insulation would have to be installed inside, slightly decreasing the size of the rooms.

In areas where the external appearance can be changed, the best approach is to resurface the outside of buildings either with 2 to 4 inches of plastic cladding or with half an inch to an inch or so of the reflective foil-faced insulation described above. The surface can then be coated to provide an appropriate color or finish. Old apartment blocks are particularly suitable for this type of treatment; although it doesn't produce a Passivhaus level of insulation, it will very substantially reduce heat losses. It will also reduce the condensation problems that arise when the water vapor in the heated internal air meets the poorly insulated external wall. These are lessons that the Passivhaus movement has learned in Germany and elsewhere. The importance of extraordinary attention to detail to ensure complete airtightness and an absence of "bridges" that conduct heat to the outside world is increasingly clear. Extreme care needs to be taken both in designing refurbishments and in actually carrying out the work. In most countries, the home construction industry is still some way from fully absorbing this point.

The German government, however, recognizes the importance of eco-refurbishments, releasing a recent statement that said the renovation of existing buildings is a central element in Germany's national climate protection strategy. Since a quarter of all energy use is in residential properties, this stance is a logical one that other countries ought to share. The intention behind the policy is to improve the energy efficiency of the housing stock by 3 percent a year. It is arithmetically obvious that this cannot come from new buildings, which each year form at most 1 percent of the total number of houses in Germany. The bulk of the improvement will therefore come from refurbishing the oldest houses, particularly those built before energy-efficiency rules on new buildings were imposed in the late 1970s.

The primary mechanism the German government is using is a subsidized long-term loan scheme for any owner wishing to reduce the energy use of his or her building. The Germans say that pre-1984 homes have three times the energy use of buildings constructed to today's energy-efficiency regulations. Typically, a refurbishment aimed at reducing energy use achieves only half of the savings that are possible. But a really effective eco-refurbishment can reduce energy use by over 80 percent, taking the energy consumption of many older homes well below the standards currently demanded for new buildings. It is a mistake to assume that only building new homes can give us good insulation performance.

The German Energy Agency started a project in 2003 to refurbish 140 buildings to prove this point to a skeptical construction industry. Many of these buildings contain multiple apartments, and in total, the study looked at over two thousand separate homes. The agency claims that the refurbishments have cut energy use so much that these buildings now have electricity and heat consumption of less than half what would be expected in a conventional new-build.

The Passivhaus Institute agrees about the potential for refurbishments. It says that it is possible to achieve energy-efficiency results in an eco-refit that are not far behind the achievements of the best new Passivhaus homes. The institute quotes a figure for the energy used in heating of about 2.3 kilowatt-hours per square foot for the best refurbishments, compared with 1.4 kilowatt-hours for a Passivhaus. For an average thirty-year-old home, getting down to 2.3 kilowatt-hours per square foot would mean a savings of about 3 tons of carbon dioxide each year.

One astonishingly successful renovation in Linz, Austria, should convince us all that converting old apartment blocks is a much better way of spending money and offers more potential savings than just insisting on ever-higher standards in new building. This fifty-unit apartment block was built in 1957 and had heating energy use of about 16.7 kilowatt-hours per square foot. The refurbishment a few years ago reduced this figure to below 1.4 kilowatt-hours. The other advantages included increasing the internal size of the properties by over 10 percent by walling in unused open balconies. The renovation also significantly reduced the traffic noise inside the apartments. Air quality in the homes is better, as incoming air, polluted by vehicle fumes, is now filtered. The cost of this groundbreaking eco-renovation was nearly

147

$75,000 per apartment, so the reductions in energy use are only likely to pay back the investment over several decades, but the improvement in the appearance and livability of the block is striking.

If improvements of this size were delivered across the whole housing stock, the prospective savings would be at least 10 percent of national emissions and possibly much more. A very welcome by-product would be a reduction in the number of people suffering from cold and discomfort in winter.

A variety of approaches were used in the 140 buildings of the crucial German study, with some of the refurbishments making extensive use of highly innovative Passivhaus components for insulation and airtightness. The results have helped shape the subsidized loan scheme now available to the general public. The German national and state governments offer a substantial incentive for achieving really good energy performance. If a renovation achieves energy use 30 percent below the building regulations for new-builds, the government will forgive 12.5 percent of the money owed. Smaller energy savings are rewarded with smaller, but still significant, savings.

All told, the German government is plowing over $7 billion per year into the eco-refurbishment of existing houses in the form of loans and subsidies. This amount seems large, but one estimate puts the total amount of money spent on all types of structural housing improvements at nearly $120 billion in Germany alone. The eco-loan scheme has been popular with landlords wanting to maintain the standards of their stock of buildings, particularly those operating in the social housing sector. In 2007, the scheme helped reduce energy use in about 200,000 homes, or about 0.5 percent of all the houses and apartments in the country. This number is almost as great as the total number of new housing units being built each year. So, unlike Britain and the other countries that are focused on regulating the efficiency standards of new buildings, Germany is seeing far greater carbon dioxide cuts from putting money into the oldest portion of the housing stock.

The sponsoring government ministry says that each year of the program has saved annual emissions of about 1 million tons of carbon dioxide, or about 5 tons per housing unit renovated.

Does the German scheme make financial sense, both to the homeowner and to the government? On the basis of the numbers released by the government, the answer is a cautious yes. If we assume the refurbishment lasts fifty years, it implies a cost of $30 for each ton of carbon dioxide saved, a highly competitive figure. And since this money is not a direct subsidy, but simply a loan, the effective cost is far, far smaller.

The impact on the homeowner is less clear-cut. The savings in fuel bills arising from the refurbishments are probably between $750 and $1,500 a year, and often less. An investment of up to $75,000 to achieve these cost reductions is not an obviously successful investment with an annual savings return of perhaps only 2 percent. (The central government is more optimistic, quoting typical returns of several times this level, but their figures seem unusually high.) Nevertheless, what is absolutely clear is that landlords and homeowners are extremely keen to carry out energy-efficiency refurbishments, whatever the short-term benefit in reducing gas and electricity bills. Refurbishment makes the home more comfortable and, for owner-occupiers, much easier to sell.

A secondary benefit to the German economy has been an upsurge in the number of jobs in the construction industry. Germany has suffered from high unemployment, particularly in the states of the former East Germany, so this side-effect has been extremely welcome. The government quotes figures of more than 200,000 people pulled into permanent employment as a result of its refurbishment program. A U.S. study by the Center for American Progress in September 2008 suggested that a strong green stimulus might replace all of the 800,000 jobs lost in construction in the previous two years, or about the same percentage of the working population as in Germany.

About three-quarters of German housing was built before the

end of the 1970s, when insulation standards began to improve. To refurbish all these houses within the next thirty years would mean tripling the current rate of refurbishment. That sounds ambitious, but the eventual savings may approach 100 million tons a year of carbon dioxide—or about 10 percent of the German total.

The technologies for improving the energy consumption of domestic housing are simple and relatively well understood, partly as a result of the groundbreaking work of Wolfgang Feist and the Passivhaus Institute. Alongside more exciting and more glamorous techniques for carbon reduction, the world needs to devote efforts to making the easy gains in house insulation and airtightness. So far, developed countries have been slow to see the potential and cost-effectiveness of refurbishment. But as energy costs rise and more and more families struggle to meet heating bills, reducing energy use in homes is an obvious and attractive means of massively decreasing emissions. And we could start tomorrow.

ELECTRIC CARS

The inevitable switch to battery propulsion

IN OCTOBER 2007, the Israeli-born Californian software engineer Shai Agassi announced a plan to accelerate the move to the all-electric car. Agassi is a visionary with a capacity to inspire change. But the scale of the task he took on is breathtaking. At the moment, car batteries for electric-only vehicles are extremely heavy, recharge slowly, have a limited range, and are eye-wateringly expensive. Agassi's scheme requires carmakers to build a new generation of all-electric cars, probably with no backup engine. His company will then offer to rent batteries to the car owners at a cost that beats the price of gasoline. Owners will recharge the cars at home or at work. For long journeys, they will simply swap the batteries at automated recharging stations conveniently situated on major routes.

Agassi has taken on one of the most important challenges we face: the need to create a world vehicle fleet powered by an alternative to globe-warming petroleum fuels and inefficient internal combustion engines. We can't know whether his venture will succeed. But he is just one of the many people betting that gasoline and diesel will be replaced by electricity. Battery technology is likely to advance very substantially in the next decade. Price, weight, range, and recharging time will all improve. In contrast, although the price of oil has subsided since the mid-2008 peak of $140 a barrel, oil is likely to get more expensive as time passes. Electricity is now very clearly a less expensive way to run a vehicle and is likely to become even more attractive in the future.

Small numbers of all-electric cars have been around for decades. They are generally slow and have a very limited range. Perhaps 25,000 of these vehicles putter around the suburban streets and retirement communities of California and other places in the U.S. But battery vehicles that look and drive like ordinary cars will be on sale in late 2010, with mass production from manufacturers such as Chevrolet and Nissan slated for 2012. Batteries will need to be recharged with electricity, and at the moment that electricity is most definitely not net-zero. But many of the other advances suggested in this book will substantially reduce the carbon produced when electricity is generated. As time progresses, the advantages of using electricity to drive our cars can only grow.

The modern car is the product of more than a hundred years of evolution. It is extremely reliable, at least compared with its ancestors. Usually comfortable and pleasant to use, it is also much safer than it was even twenty years ago. But a century of improvements hides a surprising fact. The engines that power today's cars remain inefficient and wasteful. Automobile manufacturers spend enormous sums each year on research and development, but only about a quarter of the energy in gasoline actually gets to the wheels of the car. Over three-quarters is wasted as heat, mostly from the exhaust or through the cooling system. Diesel cars are better but still waste most of the energy in the fuel.

We will see some small improvements in internal combustion engines over the next decades. The cars hurtling round the track at a Formula 1 Grand Prix turn about a third of their fuel's energy into useful motion. The billions spent every year trying to take milliseconds off lap times will eventually result in technical advances that can be used in vehicles whose most exciting ride may be a trip to the supermarket. But internal combustion engines are never going to be parsimonious in the use of fuel. By contrast, electric motors can turn 80 percent of the power delivered by a battery into useful motion. This simple fact means that it is not a question of *whether* the electric car will take over the roads but a question of when.

The answer to this question has important implications for climate change. As more and more people around the world become rich enough to afford a car, the emissions from automobile engines will increase. There are about 600 million cars on the road today. The burgeoning Chinese and Indian middle classes might push this number up by several hundred million in the next decade alone. Cars and trucks are already responsible for about 15 percent of global emissions. Without a revolution in technology, this number will inevitably rise, even if car manufacturers surprise themselves with the improvements that they can eke out of Nikolaus Otto's 1876 designs for a four-stroke compression engine.

Of course, cars only represent part of the problem. Our roads are also used by vans and trucks that transport our food to supermarkets, our clothes to shopping centers, and our waste to landfill sites. It is more difficult to convert these vehicles to batteries. Their power needs are too great, and many of them make long-distance trips that would run down even the largest battery pack. Some commercial vehicles can be converted to fuel cells; city buses are probably the best example. Other commercial vehicles can run on second-generation ethanol as a substitute for gasoline (see Chapter 7). But large trucks making long journeys will need to use liquid fuels until we find a way of making diesel from agricultural wastes. Chapters 9 and 10 look at the ways in which we can extract carbon dioxide from the air to offset the emissions from those vehicles that we can't easily convert to electricity.

WHY ELECTRIC?

To some people, the answer to slashing transport emissions is simple. We should engineer our world so that public transport replaces private cars. We should strive to reduce the distances to our workplaces and our shops so that people need to drive less. If we localized our food production, we might be able to omit our weekly drive to the out-of-town supermarket and cut the 25 percent of all truck journeys that, in many countries, are accounted for by carrying food to shops, factories, and warehouses. On those occasions that people still need to use a car, they should have access to pools of shared small cars.

These policies are all sensible. But the unfortunate fact is that the private car is a possession of high value to many people: it provides freedom, independence, and, all too often, status. Measures to damp down car ownership are fighting against strong human desires. Yes, we can reorganize our cities to make bus travel easier and more comfortable, and we can make cycling safer and more fun. But such measures will make only a small dent in the demand for gas and diesel.

To make a bigger difference, we clearly need to produce the greenest possible cars. Various options for achieving this goal are on the table: the use of hydrogen as a fuel; smaller and more efficient conventional engines; low-carbon fuels from agricultural products (biofuels); and electric cars.

Despite the interest sometimes given to the first of these alternatives, we can quickly dismiss it. There are three enormous problems with hydrogen. First, it would require a completely new set of refueling points. Thousands of filling stations would need to be established in every country. The cost of hydrogen storage is high, not because it is dangerously explosive (in most circumstances it is not) but because it needs to be stored at high pressure. At low pressure, hydrogen occupies a substantial amount of space for each unit of usable energy it contains. So it needs to be hugely compressed so that the storage tanks do not take up too much space. High-pressure storage is expensive, and the cost of building the tanks necessary to hold the gas would be prohibitive.

The second problem with hydrogen is that it takes substantial amounts of energy to make in the first place. The easiest way to create pure hydrogen gas is to split water into its constituent atoms, oxygen and hydrogen. This process is simple, but it takes substantially more energy, usually in the form of electricity, than we can usefully capture when we later burn the hydrogen in a car engine. This lack of any energy benefit is an unchangeable law of chemistry, and nothing is going to get around the problem. Of course, the electrical energy we use to make the hydrogen could come from renewable sources, in which case there would be no carbon dioxide cost. But we could have used this electricity for other purposes instead. Making hydrogen with wind power means that we cannot use the electricity to power homes and offices, or indeed to refuel car batteries. A kilowatt-hour of electricity in a battery will drive a car farther than the same amount of electricity used to make a tank of hydrogen.

Third, cars that use hydrogen as their power source only exist in very limited numbers. The Honda FCX (Fuel Cell eXperimental) is on sale in some parts of southern California that are within easy reach of the small number of hydrogen refueling stations. A few other prototype hydrogen cars are also on the roads. Manufacturers have taken two different approaches: the hydrogen can be burnt to capture energy, much as in a conventional engine, or, as in the Honda FXC, the hydrogen can be used to generate electricity in a fuel cell inside the car. As Chapter 4 explains, fuel cells are a very promising technology for static heat and power generation, but they're currently expensive and too large for use in cars.

One estimate is that each of the Honda fuel cell cars costs almost a million dollars to make. They are also not obviously suited to the vibration and rough treatment meted out to an engine as it travels over uneven roads. Fuel cell vehicles exist, but the technology is probably better suited to sedate urban buses than cars that need travel a long way from their home refueling point.

The second possible route forward is to make lower-emission cars. Governments around the world are focusing on encouraging manufacturers to build smaller, lighter, and more efficient cars. They are applying pressure both by setting maximum limits on the typical emissions of each car company, averaged across all their models, and by increasing taxation on the vehicles that consume the most gas, either through fuel taxes paid on each gallon, or by increasing excise duties on bigger cars. Proposals from the European Union to mandate further improvements in emissions standards have elicited howls of pain from the manufacturers of big and heavy cars, such as Mercedes-Benz, which will have to make disproportionately large reductions. In late 2007, the U.S. Congress similarly approved emissions targets for automobile manufacturers, which will force them to make substantial improvements in fuel economy. This move follows thirty years of gradually decreasing fuel economy in U.S. cars.

The problem is that even the smallest, lightest, most aerodynamic gas-powered European or Japanese cars emit around 6 ounces of carbon dioxide per mile traveled. Given that the average North American car travels about 13,000 miles annually, even the best new car will typically emit over 2 tons of carbon dioxide a year. Given that the world will probably need to cut total emissions across all activities to no more than 1 or 2 tons per person by no later than 2050, cars powered by internal combustion engines represent an important target for carbon reduction.

Governments have also been keen on biofuels, the third option for cutting car emissions. Carbohydrate crops, such as wheat, corn, or sugar beet, are distilled into ethanol, and oil-bearing products, such as canola seeds or the berries of the tropical plant jatropha, are turned into diesel. As the following chapter explains, however, official enthusiasm for vehicle fuels from crops is being quickly eroded by the increasing evidence that these fuels do little to cut emissions.

The next chapter looks in detail at second-generation biofuels made from agricultural and forestry wastes. The scope for using these fuels for reducing emissions is much greater than with today's crop-based ethanol, and there is considerable reason for optimism about the long-term potential of low-carbon liquid fuels. But even the new cellulosic ethanol fuels need truly enormous amounts of feedstock from forest and fields. We may never be able to fuel our needs for personal transport without causing further problems of deforestation and the loss of food-producing land. It makes good sense for policy-makers to encourage car owners to use electric vehicles. Although the move to a car fleet that is very largely powered by electric batteries has many challenges, it will eventually reduce driving costs and substantially cut carbon dioxide emissions.

Some automobiles that can switch between electric batteries and conventional gas engines are already on the road. The best known of these "hybrids" is the Toyota Prius, a good-looking medium-sized sedan. The Prius's batteries power the vehicle around cities. On longer journeys, the gas engine kicks in, and the car operates as a conventional motor vehicle. The car's batteries are recharged by recapturing the energy the vehicle loses as it brakes. Think of it this way: a car traveling at 30 miles per hour has a lot of kinetic energy. As it slows to a stop at a traffic light, this kinetic energy is lost. The law of conservation of energy says that the energy of the moving car must be translated into heat or some other equivalent energy carrier. Most vehicles turn deceleration into heat—this is why brakes get hot. A Prius is different. It captures some of the energy from braking and turns it into electrical energy in a battery, a process known

157

as "regenerative braking." The same principle is now also used in some trains and a small number of other cars.

The Prius's combination of internal combustion engine and battery has several other advantages. First, the battery can be used to move the car forward after it has stopped at intersections or traffic lights, meaning that the gas engine can be turned off when the car is stationary, rather than wasting fuel idling. Second, the battery can work in tandem with the engine to increase the power of the car when it is accelerating. This means that the combustion engine does not have to be as large as it would normally be in a car of this size. Since small gas engines deliver inherently better fuel economy than larger ones, this feature reduces the average gas consumption of the car.

As a result of having the battery available, the Prius's fuel economy was outstandingly good when it came out five or so years ago. Its carbon emissions were lower than any standard car on the road. But other manufacturers have made their own improvements, and several small cars now have fuel consumption that nearly matches that of the Prius. The Toyota iQ, a small car sold in Europe, for example, has carbon emissions of 99 grams per kilometer, only slightly more than the 89 grams of the Prius. The iQ doesn't do anything sophisticated like capture kinetic energy and turn it into electricity. It is simply a highly aerodynamic and light car that requires limited power, and its small engine is good at turning the chemical energy in fossil fuel into movement.

The Toyota Prius and other hybrid cars are the first steps on the road to fully electric vehicles. They have batteries that can only carry the vehicle for a short distance, and their electrical propulsion system has to operate alongside the conventional internal combustion engine. This makes the car expensive, complex, and heavy. The battery of the Prius weighs about 125 pounds and stores less than 1 kilowatt-hour of energy, approximately enough to drive the car at a constant speed for about 6 miles. Put another way, the battery is only able to accelerate the car from standstill to 60 miles per hour four times before it needs recharging.

Within a few years of Toyota bringing out the Prius, "hackers" had spotted an obvious improvement to the car. They added an informal modification that allowed the owner to plug the car into an electrical outlet to recharge the battery from the grid rather than through regenerative braking. These people have turned their Prius cars from being ordinary hybrids to being "plug-in hybrid electric vehicles," or PHEVs. The owners can plug the car in at night and use the batteries to drive short distances to their job every day. On longer trips, they can still use the gas engine to power the car when the batteries run down.

Many people think that PHEVs are the car of the future because they combine the advantages of battery cars and conventional engines. A very large fraction of all car journeys cover only short distances, so for many people, a car such as this would almost exclusively be powered by electricity. The internal combustion engine will kick in during the rare longer trips. The owner will have the certainty of knowing that the car will have a source of power even on an unexpected trip across the country.

But PHEVs also have disadvantages. The car's batteries are heavy and expensive. They have to fit in the car alongside the gas engine and its transmission system. Putting both an internal combustion engine and an electric propulsion into the car is very costly and reduces storage space. The pioneering Massachusetts battery company A123Systems is selling an additional battery pack for the Prius that makes it far closer to being an all-electric car. At $10,000, it's an expensive add-on, but for regular users, commuting every day for 50 miles, the battery may just about make financial sense.

A battery-only car could, in theory, be a much better idea than a hybrid like the Prius. Although the batteries would have to be even larger than in a PHEV, the cost and space savings from not having an internal combustion engine could be very substantial.

159

An all-electric car might simply have electric motors driving each wheel, and almost all of the complex system of gearboxes, cooling systems, and power transmission would disappear. As battery technology improves, building an electric car will eventually become much cheaper than even the simplest fossil fuel alternative. The only surprising thing is that this hasn't happened already.

The other advantage, of course, is that electricity is much cheaper than gas, particularly in countries with punitive fuel taxation, such as most European states. The electric car is able to convert a large fraction of all of the energy in a battery into motion, compared with about 20 to 25 percent for a gas-powered equivalent. A light electric car traveling at 40 miles per hour uses about 7 kilowatts of power. At typical Toronto electricity prices in late 2009, this power would have cost about 55 cents an hour. Even a very fuel-efficient small gas-powered car would cost over six times that amount at Ontario gasoline prices of about CAD$1 a liter, roughly US$3.50 a gallon.

An electric car is even more advantageous in countries with high motor fuel taxes, such as most countries in Europe. But in almost all countries, an electric power source will cut motoring costs by over three-quarters. The relative simplicity of all-electric cars will eventually also mean reduced maintenance and insurance costs.

So from the car owner's point of view, electricity is better than fossil fuels such as diesel and gas. And although electricity may get more expensive because of rising fossil fuel prices and the expensive subsidies given to renewable electricity generators, it is likely to remain considerably cheaper than liquid hydrocarbon fuels, even when full-scale production of next-generation biofuels begins in a few years.

What about the impact of electric cars on carbon emissions? Electricity from the grid is most definitely not net-zero. In the U.K., running a 7-kilowatt electric car for an hour will produce about 7 pounds of carbon dioxide from power stations when the battery is recharged. This figure will be higher or lower in other countries, depending on how they produce their electricity. In a country like the U.S., which generates most of its electricity from coal, an electric car powered from the grid would produce more carbon dioxide than in France, which gets most of its power from nuclear stations, or Canada, which generates much of its power in hydroelectric power plants.

An efficient gas-powered car, by comparison, would probably produce about 22 pounds of carbon dioxide if driven at 40 miles per hour for an hour. A hybrid wouldn't be much better. So an electric car has emissions of less than a third of a new gas-powered car on the road in countries using fossil fuel to generate most of their electricity. As industrial countries gradually increase the percentage of electricity coming from renewable sources that do not produce carbon dioxide, the advantages of powering our transport fleets with batteries will become even more obvious. The financial and ecological arguments for using electricity to power our cars are overwhelming. And there are other advantages. Electric cars produce no pollutants when driven around, so the air quality in cities would be much better, helping reduce the incidence of asthma and other respiratory diseases. Most electric vehicles are also very quiet, so a large-scale switch would reduce the noise level around busy roads. The evidence that high noise levels affect human health is not yet as convincing as the link between city center traffic and asthma. Nevertheless, the standard of living of millions of people would be improved if we reduced the background level of traffic noise in industrial countries.

THE ROADBLOCKS

I suspect that eventually almost all our cars will be electric. But getting from a position in which almost all vehicles are powered by grossly inefficient internal combustion engines to one in which lighter cars glide noiselessly around our cities is not going to be easy.

We face four main problems. First, batteries are expensive. Today, a power pack that will drive a car for a hundred miles will cost over $12,000. This might double the construction cost of a small family car. Second, the batteries weigh at least 450 pounds, adding 20 percent or more to the weight of a small vehicle. The time needed to recharge a battery is declining, but it can still take eight or ten hours if charged at home. And, lastly, batteries for cars will probably be based on a light metal called lithium, which is only mined in a small number of places around the world, mostly in South America. Lithium is the key element in today's batteries for mobile phones and laptop computers. Although there is no current shortage of the metal, a substantial shift to electric cars would increase worldwide demand by a multiple of a hundred. If we need to make tens of millions of car batteries every year, there will be inevitable problems getting reliable supplies at a reasonable price.

There are other minor obstacles. One problem is that it is surprisingly difficult to tell what percentage of a battery's charge remains. Users will be understandably cautious about adopting the electric car if they don't know how far they can drive without topping up their electricity. In the past, batteries would typically fail after a relatively small number of charges. Even today, lithium-based laptop batteries often cease to work some time before the rest of the computer fails. But recent improvements should mean that new batteries will last at least as long as the cars they're found in, even if they are charged and completely discharged every day.

There's another potential obstacle that rarely gets noted. Governments in Europe and elsewhere get a large fraction of their tax revenue from fuel duties and vehicle taxes. In the U.K., for example, about 6 percent of all government revenue is generated this way. Will governments have the courage to lose such a valuable source of revenue, or will they quietly discourage electric car use? The problem is likely to bite first in London, where the city's revenue is boosted by the tax charged on vehicles entering the central area. At the moment, electric cars go free and, in some places in the city, can even be recharged at no cost to the driver. Will the inevitable rise of the electric car be held back by London mayors cautious about the impact on their revenue? As the number of electric cars grows, will they be able to avoid the temptation to start taxing them? History does not inspire confidence that national and local governments will encourage technology—however green—that cuts sharply into their tax base. But because electricity is so much cheaper than gasoline, there will probably always be an incentive to switch to a battery-powered car.

What about acceleration and top speed? The conventional supposition—that electric cars are necessarily slow and sluggish—is wrong. New types of battery can deliver explosive amounts of power. The Tesla Roadster, an electric sports car, finally went on public sale late in 2008. The car accelerates from standstill to 60 miles per hour in under 4 seconds, with no noise and no gearbox strain. The Roadster is also very fast: it has a top speed that has been electronically limited to 125 miles per hour. Designed and partly assembled by Lotus Cars in the U.K., the Roadster is an astonishing piece of automobile engineering that has helped change the image of electric cars around the world. The price tag is over $100,000, but drivers will get a car with a performance that matches the best gas-powered competitors. Incidentally, the car has a range between recharges of over 240 miles, largely because the manufacturer has opted to install batteries of very high capacity. This is one of the primary reasons the car is so expensive.

At over $100,000, the Roadster is not a car for the ordinary family. How will the large automobile manufacturers around the world get to the point where electric cars become the standard offering in their showrooms?

THE ROUTE TO THE ELECTRIC FAMILY CAR

California is the center of the electric car industry. But even there, buyers face a limited choice. Those who cannot afford the Tesla Roadster or who are a little too sedate to fully use its extraordinary acceleration are left with vehicles that barely reach 25 miles per hour and are only really usable for dawdling to the local shops on quiet side roads. Their owners are happy to live with their limitations, but these electric runabouts are not real substitutes for the flexible gas-powered sedan. General Motors tried to introduce an electric car in California in the late 1990s, but its attempt failed, largely because of the high cost of the car and the limited range of its first-generation batteries. Many people still say that the demise of this car, the EV1, was hastened by the opposition of the oil industry, but the unfortunate reality was that this vehicle was simply not good enough—it didn't have the performance of the Roadster or the simplicity of the glorified electric golf carts that trundle around some California communities.

The second major market for electric cars is in England. Fuel taxation is high by international standards, so electric propulsion looks like particularly good value. Car owners also have to pay a yearly tax on their vehicle, but low-carbon cars pay little or nothing. Perhaps most importantly, the "congestion charge" imposed on vehicles entering and leaving central London exempts electric cars. Unsurprisingly, then, the world's small band of electric car manufacturers has made the city a focus for their sales efforts. The Indian manufacturer of the G-Wiz has sold over a thousand of its spectacularly ugly electric cars to London's commuters. Until recently, this manufacturer used the old-fashioned lead-acid battery (the type that powers the starter motors in gas cars), and this limits the range of the car to a few tens of miles. Top speed is theoretically more than 40 miles per hour, though it can be much lower when going up London's few hills.

Such limitations may not matter much in central London, where driving distances are short and, as the company selling the G-Wiz points out, travel speeds barely exceed walking pace, even if you are in a Tesla Roadster. But for the average driver outside congested city centers, speed, acceleration, and range do matter. Thankfully, the technology is improving fast, and it should soon be possible to power a reasonably sized car for a long commuter journey with batteries that aren't prohibitively expensive and that can comfortably accelerate the vehicle to highway speeds. Early lead-acid batteries had neither the power nor the storage capacity to do this, and neither did the second-generation nickel-metal hydride technology found in the Toyota Prius. Only with new lithium-ion batteries have good range and power been possible.

Unfortunately, current lithium-ion technology has two significant problems—expense and a propensity to explode if improperly manufactured or mechanically abused. The chemistry in lithium-ion cells means that the stored energy can rapidly be released in the form of heat under certain circumstances. In 2006, manufacturing flaws that left some impurities in batteries for laptops meant that a small number caught fire, and many tens of thousands were recalled by the manufacturers. When fully charged, lithium-ion batteries in cars might contain a thousand times the energy stored in a laptop. So safety measures have to be effective and reliable.

The current consensus is that this small risk of explosion may mean that first-generation lithium-ion cells are not the most appropriate technology for mass-market cars. An alternative, usually known as "lithium iron phosphate" (LiFePO4), may well be the future choice, but a variety of similar technologies are fighting for market acceptance. Compared with their lithium-ion predecessors, LiFePO4 batteries are slightly heavier for each unit of power that they contain. But they will not explode or catch fire, they can be charged quickly, and they will work for thousands of charging and discharging cycles. LiFePO4 will eventually be cheaper than conventional lithium-ion, too, partly because the new technology uses iron, an inexpensive metal, where lithium-ion batteries require expensive cobalt.

Nanotechnology, the science of materials at extremely small scales, should enable future batteries to charge more quickly. By changing the structure of the anode of the battery almost at the level of individual molecules, we will likely eventually be able to recharge a car battery in minutes rather than hours. This improvement has not yet gotten far beyond the university laboratory, and it remains to be seen whether the technology will ever be cheap enough for the mass market, but it's possible that within five years we will be recharging car batteries almost as fast as we fill up a gas tank today.

Unfortunately, technological advances will not necessarily improve the storage capacity of batteries. There are real and unbreakable rules about how much energy each pound of each type of battery can hold. We know that if $LiFePO_4$ continues to be the technology of choice, then we will always need approximately 13 pounds for every kilowatt-hour of electric power. A kilowatt-hour will take a standard small car about 4 to 5 miles, depending on its weight and the amount of acceleration used. So for every mile of range, we will need about 3 pounds of battery. There's no reason to suppose that we won't develop lighter batteries in the future, but there are no obvious breakthroughs on the horizon.

Automobile manufacturers and battery makers face a difficult task in finding the right balance. Do they put heavy batteries in an electric car, increasing its range? Or do they reduce the weight, keeping the battery cost down and adding to the space for passengers and luggage, thereby limiting the distance the car can travel between recharges?

Perhaps the answer is that the car companies will offer two types of vehicle. Some models will be sold with smaller batteries that enable the car to be driven to work, recharged there, and then driven home. After all, the typical American commuter journey is about 25 minutes. So for most car owners, a battery that lasts, say, 40 miles is going to be perfectly adequate for most journeys. If they need to travel farther, they can borrow a car with a greater range from the company car pool or a local car-sharing club.

The optimists in the industry believe that the cost of batteries for a car of this type will fall to about $150 per kilowatt-hour in the next decade, or somewhat less than $1,500 for a range of 50 miles. This is at least a four- or five-fold reduction. There will also be a cost for the electronics needed to control the battery's charging cycles. The manufacturer will save by not having to include an engine and a transmission system in the car, but an electric car will likely cost somewhat more than a gas-powered equivalent for some time. But since the yearly running costs will be a half or a third of existing levels, the car's price tag isn't necessarily a major obstacle.

Users who frequently drive longer distances will need a car with more battery capacity or will decide to buy a hybrid vehicle. General Motors has decided that the way forward is to add a small internal combustion engine to its Volt electric car, due to be launched in late 2010. When the battery is running down, the vehicle's 1.0-liter gas engine will start and begin to provide electric power to the motor. The engine will never directly drive the car, unlike with the Toyota Prius, which switches between electric and gas propulsion. GM's solution—giving the engine the simple role of onboard electricity generation—seems simpler and more elegant than the traditional hybrid design and should help minimize production costs. The battery pack in this vehicle is quite small—only 16 kilowatt-hours—and the manufacturer says that it will only be allowed to run down to 30 percent charge. This means its range will only be about 40 miles, or enough for about 70 percent of daily commuting journeys in the U.S. The price is higher than gas-driven equivalents, but the cost is reduced by a $7,500 government subsidy.

167

Nissan also intends to launch an electric car in 2010. Its LEAF model will aim for 100 miles between charges, with the battery containing about 24 kilowatt-hours. The company has talked about leasing the battery to users for a fee of about $150 a month, which seems like a lot, but the net cost will be lower than most people spend on gasoline. Nissan's 2010 cars will be manufactured in Japan, but as full production ramps up, the Tennessee factory will start also assembling the vehicle.

And what about the scheme Shai Agassi proposed, described in the first paragraphs of this chapter? Agassi's insight is that users will be put off by the higher initial costs of electric cars but would be willing to pay to rent the batteries out of the fuel savings that they make. His scheme requires car manufacturers to design vehicles that offer easy swapping of discharged batteries with freshly charged replacements. Renault-Nissan has committed to working with his company. He also needs to persuade local entrepreneurs or governments to install a network of swapping points that give long-distance drivers the security of knowing that they can get new batteries when they need them. This challenge is enormous, and, given its huge size, the U.S. may not be the best place to start. Instead, Agassi seems to have persuaded the Israeli and Danish governments to back his scheme. The smaller size and shorter travel distances of these countries make the goal more achievable. Renault-Nissan has promised to make at least 100,000 electric cars available by 2016 in these two countries to provide a market for the new charging stations and battery-swapping points. Israel's central position in Middle Eastern conflicts makes the country unusually vulnerable to loss of its oil supply, so the government has a strong incentive to support Agassi's vision over the long term.

In Denmark, Agassi's partner is DONG Energy, which runs many of the offshore wind farms in the country. DONG will supply renewable energy for the batteries, meaning that the owners of the cars will know that their driving is genuinely low carbon.

On those occasions when DONG is generating too much energy to be used on the Danish grid, it will also make obvious financial sense to use the spare capacity to charge car batteries. Similarly, Agassi's Israeli partners are intending to use solar energy from the Negev Desert.

Meanwhile, the Norwegian company Think has launched its all-electric city car in Europe. This vehicle has a range of over 120 miles and a top speed of 60 miles per hour. The batteries can be completely recharged overnight. Eager to show that it is a "real" car, the company has constructed the vehicle with a steel cage to meet international safety standards. This is no electric golf cart. Even though its acceleration leaves a lot to be desired, this car can compete against gas-powered equivalents. Early reviews in the U.K. were a little patronizing, focusing on some of the less-sophisticated features of the vehicle, but Think is delivering a product very close to the performance of a conventional compact car. At over $20,000 in the U.K., the price is somewhat higher than the equivalent car with a conventional engine, and costs will need to come down sharply if it is to achieve large numbers of sales. But since the Think car qualifies for exemption from the London driving tax, and driving into London every working day costs over $3,000 a year, some people will be prepared to pay the hefty price for the car.

Cars like the Think demonstrate the opportunities and threats to Agassi's and Renault-Nissan's plan. On the one hand, by separating the ownership of the car and the car's batteries, Agassi's business could reduce the purchase price of an electric car substantially. But if my earlier prediction is correct and we eventually see battery recharging times falling to five or ten minutes, then Agassi's scheme will fail. It won't make sense to design cars to have easily removable batteries and to install heavy machinery at hundreds of locations to take them out and automatically replace them. If we can completely charge a car battery in five minutes, we will simply get used to driving into a "filling" station, plugging the car in, and having a coffee. And if electric cars remain too expensive because of the up-front cost of the battery, then car-leasing companies will come forward to offer better financing terms. The world may not need Shai Agassi's scheme for kick-starting the electric revolution after all.

CARS AND THE GRID

The way to battery-based driving will encounter substantial obstacles, but the economic and environmental arguments are too compelling for the electric car not to eventually win the day. Electric cars will save drivers money and, with private cars and light commercial vehicles responsible for 20 percent of European carbon dioxide emissions and more in the U.S., they represent a big step forward on the road to a low-carbon future.

Even if vehicles are charged by electricity made at a coal-burning power plant without carbon capture, they will save emissions. But that, of course, is not the vision. We want cars to be charged by electricity from, and to work symbiotically with, renewable power sources. The proposed joint venture between Shai Agassi and DONG Energy in Denmark is a good example of this relationship. In the longer term, we want car batteries to act as electricity storage for the grid. Renewable electricity can be intermittent (as with tidal energy) or both intermittent and unpredictable (as with wind). As described in Chapter 1, car batteries linked to the grid with intelligent communications could stop charging when supply falls and even feed energy back into the grid when necessary.

A million electric cars parked at homes or offices, and plugged into the grid, could meet 5 percent or more of the electricity supply in a big country such as Germany, possibly within a few seconds. This might double the emergency buffer of electric power held by the operators of national electricity grids. The value of this system would be enormous, both in allowing the country to use more intermittent and unreliable energy sources and in reducing the need for fossil fuel power plants to be kept on standby. It might also significantly improve the attractiveness of owning an electric car, because utilities will be prepared to pay for the value of the storage capacity in the battery. (Rather optimistic estimates produced by Hillary Clinton during the 2008 U.S. primary campaign even suggested that the financial value to the utilities of having access to the batteries would be greater than the cost of the batteries themselves. But if this were true, it would make sense today for the electricity companies to be buying their own batteries.)

Turning car batteries into emergency stores of power is all well and good, but electric cars are still ultimately consumers of electricity. So will a switch to battery-powered vehicles put an impossible extra demand on power generators? Shai Agassi says that if every car in Israel were powered by electricity, it would only add 6 percent to national electricity demand. Rough calculations suggest that the figure might be about 12 or 15 percent in the U.S. These numbers show that the extra electricity use can easily be met from renewable sources such as solar power in Israel, wind in Denmark, and tidal or wave power along the northwest coast of North America. With intelligent charging systems, we will only be sending electricity to the batteries when there is a surplus in the national electricity distribution system. Recharging will mostly happen at night, when demand is relatively low and most cars are parked next to homes with easily available power sockets. When power is short, charging will cease, and the flow of electricity will be reversed as batteries help balance electricity supply and demand. Because we can be sure that charging car batteries will not add to the daily peak demand for power, we will not need to construct new power stations. However, we do need to be aware that there may be some times when it becomes impossible, or extremely expensive, to charge a car's batteries because the grid is unable to produce enough power. At these moments,

Shai Agassi's battery rental scheme will come into its own. The car owner would drive to the local battery exchange point and swap his or her partly charged units for full equivalents.

Of all the technologies in this book, battery-driven cars have advanced furthest in the last year. The flow of news during 2009 has been remarkable, with many major manufacturers announcing the development of new ranges of electric cars. By late 2010 or 2011, most vehicle manufacturers will offer a range of plug-in hybrids or even electric-only cars. Optimism should be tempered by the fact that hybrid cars were still only 3 percent of U.S. sales in fall 2009, perhaps because of the large price premium, but electric cars are going to make rapid advances, buoyed by quickly declining battery costs and the prospects of higher prices of gasoline in the future.

The twenty or so major car manufacturers are increasingly aware that for the first time ever, their dominance of world markets faces a powerful threat. The internal combustion engine and complex mechanical transmission and braking systems of a typical car are eventually going to be considerably more expensive than simple and reliable electric cars. Despite their huge manufacturing scale, marketing skills, and worldwide network of retailers, these companies could be quickly undermined by new and fast-moving competitors such as Tesla and Think. Growing awareness of this fact has pushed the global car-manufacturing industry to change gears. After dismissing the prospects of battery cars for decades, the car manufacturers are beginning to invest in their own electric vehicle projects. This is excellent news: we need these companies, with their huge resources and skills, to get behind the electric car. As with several other technologies in this book, only the support of the very largest companies will enable us to develop low-carbon alternatives at the speed the world needs. The chief executive of Renault-Nissan, Carlos Ghosn, once said, "We must have zero-emission vehicles. Nothing else will prevent the world from exploding." Ghosn has an apparently unshakable faith in the necessity of a rapid swing away from the internal combustion engine. His partnerships with Shai Agassi, his commitment to build a mass-market small electric car, and his public pronouncements about battery technology have helped convince the world's motor industry that they too need to begin rapid development of cars that make traditional large engines redundant.

⬳⟨7⟩⬲

MOTOR FUELS FROM CELLULOSE

Second-generation biofuels

A T THE 1900 World's Fair in Paris, the Otto car company demonstrated an unmodified diesel engine running on peanut oil, not conventional fuel. Rudolf Diesel, the inventor of the engine that bears his name, noted the success of several other attempts to use food crops as alternative fuels and later wrote that "power can... be produced from the heat of the sun, which is always available for agricultural purposes, even when all natural stores of solid and liquid fuels are exhausted." One of the other great figures in the early history of the motor car had a similar view. In 1925, Henry Ford said:

> The fuel of the future is going to come from fruit like that sumach [a type of tree] out by the road, or from apples, weeds, sawdust—almost anything. There is fuel in every bit of vegetable matter that can be fermented. There's enough alcohol in one year's yield of an acre of potatoes to drive the machinery necessary to cultivate the fields for a hundred years.

Though few people realize it today, plant-based biofuels, as described by Diesel and Ford, were real competitors to petroleum fuels in the early decades of the car. Early Ford automobiles could run on alcohol, and several U.S. distilleries turned agricultural crops into fuel for cars until the 1930s. Only the advent of Prohibition finally stopped the manufacture of alcohol as a fuel, clearing the way for the hegemony of crude oil.

The oil price shock of the early 1970s saw a temporary revival of interest in use of agricultural crops as a source of motor fuel. More recently, the U.S. has strongly encouraged the use of corn as a source for gasoline. Government policy was initially driven by a wish to provide new markets for corn farmers, who had been suffering from declining incomes. Only in the last four or five years has ethanol been seen as a way of reducing the U.S. dependence on oil imports and addressing the climate change problem. From a global warming perspective, food crops are potentially better than fossil fuels because burning ethanol simply returns carbon to the atmosphere that had previously been extracted by photosynthesis when the plant was growing.

Today, many countries have policies to increase the use of crops as fuel. A large fraction of Brazil's fuel is made from sugarcane. Most crucially, about one in twenty of the world's cereal grains is now processed by U.S. refineries into a gasoline substitute. To put this in context, over 100 million tons of North American corn is turned into biofuel each year, but this only shaves about 1 percent off world oil demand.

Unsurprisingly, turning huge quantities of corn into fuel has tightened the world market for foods. Prices rose dramatically in 2007–2008 and at the time of writing are still well above levels of five years ago. One International Monetary Fund survey indicated that the use of corn for biofuels was responsible for about 70 percent of the increase in the world price for corn between 2004 and 2008. Using increasingly scarce food to make fuel has given biofuels a bad reputation. But advances in chemistry mean that we will soon be able to use agricultural wastes such as wood chips and straw to make fuel, just as Henry Ford forecast eighty years ago. The key is cellulose, a complex and tough molecule that forms a large part of almost every growing plant. Cellulose is the most abundant carbon-based molecule in the natural world, vastly more abundant than the simple sugars and carbohydrates we are now using to make biofuels. We need to find a way of cheaply and efficiently cracking the cellulose molecule and turning it into simple alcohol. There are challenges still to overcome, but with huge amounts of U.S. venture capital flooding into the industry, cellulosic alcohol may well become the liquid fuel of the future. Combined with electric cars, cheap and environmentally benign ethanol can help slash carbon emissions from transport.

FROM BAD ETHANOL TO GOOD

Add some yeast to sugary liquids in the absence of air, and the resulting fermentation process will produce alcohol—called ethanol if you use it as a gasoline substitute. Ethanol burns well and can be added to gas as a supplementary fuel for today's cars. In fact, with some slight modifications, many cars on the road today can run perfectly well on almost pure ethanol mixed with just a small amount of gasoline. In some ways, this fuel is actually better for cars than gasoline. It delivers better acceleration and reduces the need for potentially dangerous additives. And although ethanol contains less energy in each gallon than gas does, new engines with high-compression cylinders may be able to turn slightly more of the fuel's energy into motive power than is the case with gas-powered cars.   177

Most of the biofuel sold in Europe and America today is made from foodstuffs, such as sugarcane, wheat, and corn, which have been turned into simple sugars and then fermented by yeast. By contrast, biodiesel is made by crushing seeds to capture the natural oils they contain. Diesel's engine exhibited at the Paris World's Fair used peanut oil, but today's favorite sources are tropical palm oil and canola oil from temperate areas such as northern Europe. After harvesting, the oils are put through a chemical process to create a diesel substitute. In both cases, high-value agricultural products are being diverted and taken through inefficient processes to create fuel for the ever-increasing number of cars.

Car owners filling up their tanks in Europe will be generally unaware that their expensive fuel already has a small amount of ethanol or biodiesel mixed with the fuel. As a means of getting the industry started, retailers in some parts of the world are now obligated by law to incorporate biofuels into standard gasoline. Similarly, diesel fuels have had plant oils added to the mix. When the large-scale move to biofuels began about five years ago, the world was a very different place. Agricultural surpluses were holding down the price of grains. Ethanol refineries that sprang up were welcome additions to the flat landscapes of grain- and corn-producing areas around the globe. In Brazil, the sugarcane industry now produces huge quantities of the cheapest ethanol in the world from a crop that otherwise faced continuous downward price pressure because of the export subsidies of the rich world.

As we have begun to understand the true impact of ethanol and biodiesel production, however, serious doubts have arisen about the wisdom of requiring oil companies to incorporate biofuels into their products. Most people now think that the benefits of first-generation food-based biofuels are outweighed by the problems they cause. Even the politicians who initially backed the ethanol industry are now edging toward the view that mandating gasoline to contain more ethanol would be a mistake. Food price increases in 2007–2008 prompted riots and increasing worries about security of supply. Ethanol production also uses large quantities of water in places where the supply may be already worryingly insufficient.

Furthermore, although ethanol made from crops may help reduce dependence on imported oil, it probably does very little to reduce the emissions of greenhouse gases. Growing wheat in Europe or corn in the American Midwest requires large inputs of fossil fuel energy to produce the fertilizer, look after the growing crop, and process the grain into sugars and then ethanol. Moreover, when it breaks down chemically in the soil, artificial fertilizer produces a small amount of nitrous oxide, a greenhouse gas over three hundred times as damaging as carbon dioxide.

Although the precise figure is the subject of fierce and bad-tempered disputes between scientists and fuel manufacturers, ethanol made in temperate countries probably saves less than 30 percent of the greenhouse gases associated with a similar amount of gasoline. As knowledge improves, we may find that ethanol from wheat and corn actually saves no emissions whatsoever. In particular, the evidence is growing that in some climates and soil types, growing wheat using nitrogenous fertilizer generates enough nitrous oxide to wipe out the greenhouse gas benefits of using wheat rather than gas. Despite this, European countries are still pressing ahead with a plan to ensure that at least 10 percent of motor fuels are made from plant sources by 2012. Biofuels might conceivably reduce transport emissions by a few percent in the EU, but this progress will be overwhelmed by the emissions from the extra cars on the road.

It also seems highly likely that biofuels exacerbate the problem of deforestation. Perhaps a fifth of human-made greenhouse gas emissions come from the clearing of forests. When a forest is destroyed, much of the carbon stored in its trees and soils becomes carbon dioxide in the atmosphere. As larger fractions of food-producing land are given over to ethanol and biodiesel, the pressure to cut down forests to replace the lost cropland increases. This factor is particularly important in the tropics. Old forests are being destroyed in order to plant oil palms for biodiesel in Asia. Even in Brazil, the loss of the rainforest appears to be exacerbated by ethanol production from sugarcane. Although cane is produced in the drier parts of the country, well away from the Amazon, the use of agricultural land for growing crops for fuel is affecting the supply and demand for land across the entire country. The unpalatable conclusion is now almost undeniable: biofuels made from foodstuffs are adding additional stresses to an already overstretched world ecosystem.

The core problem is that the amount of energy used to drive people around is huge, far greater than the energy in the food that we eat. The average person in the U.S. uses over 180 gallons of motor fuel a year, or about a gallon every two days. We can easily calculate the amount of energy contained in this fuel and compare it with the calorific value of the food we eat. A food calorie is just another way of expressing a unit of energy. We can't create new energy by turning wheat or corn into fuel; this would break the laws of physics. All we can do is convert the energy into a different form. Every calorie we use to make motor fuel reduces the calories available to eat.

The result of the comparison between the energy used in a family car and the calories in food is very striking and somewhat depressing. The amount of energy North Americans use to drive their cars each day is about twenty times the calorific value of the food they eat. The comparison shows the folly of any attempt to use agricultural production as a means of reducing the greenhouse gas emissions from gasoline. Even if we turned *all* our food into motor fuel and lost no energy in the process, we would only produce a tiny fraction of our total need for fuel. There simply isn't enough viable cropland to feed 6 billion people and fuel hundreds of millions of cars as well.

Comparing the energy we use driving a car with the energy in the calories from the food that we eat is not strictly fair. A lot of agricultural land—some people say 60 percent—is not actually used to produce food for human beings. Instead, it is devoted to growing food for animals, which we then eat. This process is very inefficient and wasteful: an intensively farmed cow eats 8 pounds of corn for every pound of weight in the slaughterhouse. If the world moved to a diet entirely composed of cereals, fruits, and vegetables, then there would be huge amounts of surplus land no longer needed to produce food for animals. We could use these acres to make food to convert into fuel. But even if all this land were devoted to making ethanol feedstocks, the area could only provide a small fraction of today's fuel consumption. Anyhow, the opposite is actually going on: as people get richer in the developing world, they are tending to adopt the dietary habits of rich countries and consuming more meat, thus increasing the total amount of land needed for food. The first-generation biofuels industry in the U.S. contends that increased meat consumption in the newly prosperous countries of Asia has actually had more effect on food prices than converting corn into ethanol. In addition, more and more people have access to a car, so the amount of vehicle fuel needed will continue to rise for many years.

The conflict between food and fuel can be easily shown in an example. The most productive lands in the U.K. are largely given over to wheat. In a good year, these farms can produce more grain per field than anywhere else in the world, an average yield each summer of just over 3 tons an acre. Turned into ethanol in the most efficient processing plant in the world, this might produce about 1,000 gallons of ethanol, enough to cover the annual transport needs of about six Europeans. However, the food value in the wheat would give at least fifty individuals the calories that they need for a healthy diet. Unless we can substantially expand the area given over to crops across the world, ethanol from grains is in direct competition for the limited amount of land available for growing food.

Another problem is that the current technologies used to make ethanol from wheat and other foodstuffs are not very successful at turning the energy in the crop into liquid fuels. Large amounts of heat are needed to drive the process, reducing the net energy benefit of the grains. Paradoxically, it might actually be better to burn the crop and use the combustion process to drive turbines to make electricity. The energy from a ton of corn would drive an electric car farther than the same amount of grain converted to ethanol. In the long run, as argued in Chapter 6, electricity is a far better way of powering our motor vehicles. However, we have to live with the world as it now, with over 600 million cars on the world's roads, all but a few of which run on liquid fuels.

At some point soon, the tide may turn and politicians will start to campaign against *all* biofuels. This would be an unfortunate mistake. Biofuels made from the simple starchy and sugary molecules in food are just the first stage in the exploitation of biological materials for use as gasoline and diesel replacements. The next generation of biofuels will not use the seeds of wheat and corn to make gasoline replacements; they will use the much more complicated molecules contained in wood and agricultural wastes. Eventually it will probably be possible to process any complex material containing carbon atoms—plastics, municipal wastes, even the output from sewage farms—into a liquid fuel that can be burnt in a car engine. Henry Ford understood this concept almost a century ago. The technological problems are not especially large, but dicing large carbon-based molecules into simple alcohols such as ethanol is costly and difficult to achieve on an industrial scale. We know *how* to break down most hydrocarbon molecules using some combination of heat, pressure, water, acids, and catalysts. But to make this process competitive with crude oil at a price of $70 a barrel, we need to be able to do it for less than 45 cents a gallon at volumes of millions of gallons a day. Therein lies the challenge.

Next-generation biofuels should enable us to avoid most of the problems that have arisen with wheat and corn. Many wastes, such as wood chips from sawmills, have few alternative uses, and their use for fuels will not increase the price of food. In addition, wood plantations do not generally use artificial fertilizers made from fossil fuels, nor do they require extensive cultivation from equipment burning large amounts of diesel. The same is true of grasses such as miscanthus and switchgrass, which are both very good sources of cellulose and can be grown on poor-quality land. This means that the greenhouse gas emissions from cultivating the raw materials are low. In time, we should be able to produce liquid fuels that have a relatively low environmental cost, with accompanying greenhouse gas emissions perhaps 10 or 20 percent of the impact of gasoline or diesel. This is not to say that woody biofuels will not be found to produce other environmental problems of their own, but the evidence so far is that their side-effects are minimal compared with the burdens placed on the world by ethanol made from food.

How long will it take to get to the point at which biofuels made from wood, waste, and other unusual materials can compete on price with fossil fuels? The answer may be as little as five years, but it is impossible to be certain. It could be decades. As with other technologies in this book, progress to date has been slower than expected. Vinod Khosla, the legendary Silicon Valley venture capitalist, has invested in a wide range of U.S. companies all trying to find low-cost, large-scale ways of breaking complex molecules into simple alcohol or other fuels. After a career in the computer software and networks industries, he has focused his almost limitless energy on technologies that may provide a way of cheaply converting cellulose to ethanol. But even he has found it slow going.

In an interview with Reuters in January 2007, Khosla said, "2007 will be the year cellulosic ethanol will become a real prospect for investors." He was too optimistic. Although the number of press releases announcing breakthroughs and cost reductions grew throughout the year, the evidence of real progress was small. By March 2008, Khosla was telling the *Wall Street Journal* that "the first commercial plants that are cheaper than both oil and corn ethanol are targeted to start operations at the end of next year [and will] probably be in full operation in 2010." This probably won't happen either.

So the time lines have slipped and will probably slip further. But there's little doubt that the engineering issues will be solved at some time in the next few years. The successful firm will achieve ethanol costs very similar to a gallon of gasoline. One of the eight very different firms into which Khosla has put his money will get there, but even he cannot yet know which one will achieve this ambition first.

Of course, lowering cost isn't the only issue the cellulosic ethanol industry faces. There's also the question of whether there's enough waste biological material available to fuel any sizeable proportion of the world's cars in twenty years' time. Khosla believes the answer is an unequivocal yes, though not everyone is equally optimistic. We'll discuss his analysis later in the chapter.

The previous chapter suggested that the world will eventually power most of its vehicles with electricity stored in rechargeable batteries, which is probably the most energy-efficient way of providing people with personal mobility. However, for cars still using internal combustion engines, including plug-in hybrids, we need to develop fuels that are not based on oil, coal, or gas. A new car sold today will probably still be on the road in fifteen years' time, and so we will need gasoline and diesel for at least that long and probably many decades more. How will we get to where we have substantial supplies of low-carbon ethanol that does not add to the pressure on the world's limited resources of good land?

In the U.S., first-generation corn-based ethanol is made using a two-step chemical process. First, finely ground cornmeal is added to water, and an enzyme converts the starches to dextrose, a simple sugar molecule. Then the liquid mash is transferred to a tank to which yeast is added. The yeast turns the dextrose into ethanol and carbon dioxide. Today, the most efficient ethanol refineries need about 20 pounds of cornmeal to produce a gallon of ethanol. At today's prices, this is more expensive in raw material costs than oil, and fermentation additionally requires large amounts of energy in the form of heat. So the U.S. ethanol industry only exists because of legal requirements, import restrictions, and extensive subsidies. The process is financially costly as well as being of minimal benefit to climate change. (Brazilian sugarcane ethanol is much cheaper to make because the raw material is already a simple sugar and because the heat for the process can be provided by burning wastes from the sugarcane plant itself.)

Corn and wheat have become expensive. By contrast, agricultural and forest wastes can be obtained for very little. The straw from fields, the husks of corn cobs, and the chippings from a sawmill still have little or no monetary value. However, these materials, like all plant matter, contain cellulose. Straw and leaves are mostly composed of the molecule, and even in wood it can represent over 50 percent of the weight, along with lignin and hemicellulose. Cotton has even more: about 90 percent of its mass is cellulose. Our old clothes could eventually provide a feedstock for cellulosic ethanol refineries.

Plants and trees create large carbon-based molecules, like lignin and cellulose, to provide the structure and strength they need to grow and flourish. These compounds must be solid and tough to maintain cell walls and to serve as the "skeleton" of the organisms. It is therefore no accident that they resist the attempts of chemical engineers to break them into smaller units. Whereas corn starch requires relatively little encouragement to break into sugars and then into ethanol, cellulose is very stable. We can see this in the human digestive system: foods composed of starch are easily broken down, but cellulose, commonly known as roughage, passes through the body untouched by the fierce stomach acids and hungry gut bacteria. Cellulose-digesting animals, such as cows, sheep, and other ruminants, need much more complex digestive processes and enzyme-secreting bacteria to crack the cellulose molecule in their multiple stomachs.

Cellulose is a very long and straight chain of thousands of glucose molecules bonded together. An individual cellulose molecule binds strongly to its neighbors, giving cellulose its fibrous, rope-like characteristics. These complex structures have, of course, been made by living organisms from the simple ingredients of carbon dioxide and water using the energy made available by the photosynthesis process. So we can think of the usable energy contained in cellulose as stored solar radiation. This makes cellulose a source of renewable energy, but not necessarily net-zero: we may still need energy to grow and process the chemical energy stored in the tightly bound cellulose molecules.

Scores of companies, almost all in the United States, are trying to find the best way to turn cellulose into ethanol. They're seeking a process that's cheap, that can be carried out at a scale of hundreds of millions of gallons a year, and that uses readily available sources of raw material, such as wood wastes. Most of the companies are focusing on one of two possible processes. The first option is to heat the cellulose in the absence of air to a very high temperature until it breaks into smaller molecules and eventually turns into simple gases such as hydrogen and carbon monoxide. These gases are then passed over catalysts to form ethanol, or bubbled through a stew of microbes that eat the dissolved gas and excrete ethanol. This is usually known as the "thermochemical" approach.

The number of American companies racing to find the most profitable way to produce cellulosic ethanol probably even exceeds the large number of British companies trying to commercialize energy collection from the oceans. At the time of writing, it is impossible to tell which are going to succeed and which will lose all their investors' money. If past experience in other industries is any guide, we will see one or two manufacturing technologies emerge as the lowest-cost ways of making liquid fuels. When the winning approach becomes clear, most of the original young companies funded by venture capital will quietly disappear. At the same time, larger, less nimble businesses will enter the market, hoping to compete with the successful innovators who found the best way to make fuels cheaply.

Out in front at the moment is probably Range Fuels, one of the Khosla Ventures companies. Along with five others, Range won a $76-million award from the U.S. government to help it build its first commercial-scale plant, and it has since received substantially more money from the taxpayer through the 2009 green stimulus package. The ground was broken in late 2007 for its first refinery in Soperton, about 150 miles from Atlanta, Georgia, in the middle of actively managed pine forests. The refinery will take the branches of the trees, which would otherwise be of little value, and use them to make what Range thinks will be the first commercially available cellulosic ethanol in the U.S. By the end of 2010, the company expects to have finished building the plant, a unit that will eventually turn out 20 million gallons of ethanol and other alcohols per year. When fully complete, the plant will refine about 100 million gallons each year.

These numbers are impressive, but even this large refinery, costing over $200 million, will ship less than 0.1 percent of U.S. gasoline demand. Future plants will probably be much less expensive, but these figures demonstrate the extraordinary scale of the investment that will be needed if cellulosic ethanol is to make a real dent in gasoline consumption.

Mitch Mandich, who was CEO of Range Fuels during the initial phase of the company's development and is still a director, is upbeat about the company's technology. In 2007, when construction began on the Soperton plant, he announced that his company's process yielded ten units of energy for every one put in. This makes Range's approach one of the most energy-efficient of any of the cellulosic ethanol technologies now in development. If the plant turns out to achieve this yield, it will be an extraordinary improvement on corn ethanol plants, which may use three or four times as much energy to produce each gallon of fuel.

To emphasize the wider environmental credentials of the Range Fuels approach, Mandich also said that the plant would only consume 25 percent of the water used by a refinery using corn as a feedstock. He also described how the owners of the Georgia forests in which the Soperton facility is sited plant two trees for every one cut down and how the refinery is engineered to produce virtually no waste products. If the refinery works as designed, the impact on the local environment will be very limited. This is another important way that cellulosic ethanol manufacture will be an improvement on first-generation biofuels.

The Range Fuels process is relatively simple. It uses the thermochemical route, converting solids to gases, and then the gases to liquids. The company has experimented successfully with many types of biomass, but at Soperton, the feedstock will be unused wood from the forest. The material will come into the plant as fine wood chips. Heat, pressure, and steam break the chips down in a process known as gasification. The resulting gases then react with steam to produce "syngas," a mixture of carbon monoxide and hydrogen. (This is the same process used in many fuel cells, as described in Chapter 4.) After impurities are extracted, the syngas is fed into the second phase of the process: transforming the stream of mixed gases into ethanol and other liquids with similar molecular structures, achieved by passing the gas over a catalyst, an agent that induces a chemical reaction but remains unchanged itself. Further processing then maximizes the overall yield of pure ethanol.

Which other companies look as though they might make it through the start-up phase? ZeaChem is a Colorado firm run by people with extensive experience in chemical engineering. The company's approach is to use a combination of chemical and biological processes. One stage uses a common bacterium that lives in termites and helps these wood-eating insects digest their food. This microbe turns cellulose from wood chips into acetic acid, better known as vinegar. The acetic acid goes through an intermediate stage in the ZeaChem process and is then turned into ethanol with the addition of hydrogen. The hydrogen comes from the lignin present alongside cellulose in wood and agricultural wastes. It is produced, as in the Range Fuels process, by gasifying the wood by applying intense heat. The energy for the process is provided by burning surplus hydrogen that is not needed for ethanol production.

This process currently exists only in the laboratory, but ZeaChem raised the funds for its first plant in early 2009 in Boardman, Oregon. Impressive claims for the technology include an extremely low cost of production—80 cents a gallon. Of course, this figure will only ever be achieved in a large plant when the technology has become mature. It is also dependent on finding extremely cheap bulk sources of woody material, so, as with all cellulosic ethanol producers, it's important to find raw material that the owner is willing to almost give away. For the initial plant, ZeaChem has signed an agreement with huge nearby farms that grow intensively managed and fast-growing poplar trees. Waste derived from processing the poplars will provide the feedstock. Indeed, all cellulosic ethanol producers are likely to site their plants near easily accessible forests or productive grasslands able to feed millions of tons of low-value biomass directly into the refineries with minimal transport costs.

ZeaChem, one of the few visible cellulosic ethanol companies not partly funded by Khosla Ventures, is proposing a slightly more complex process than its competitors. If it succeeds, the three main advantages of its approach are likely to be that it achieves very high levels of energy productivity, that it does not create carbon dioxide as a by-product, and that it uses almost all the products of the wood.

The claims for energy efficiency are impressive: energy output in the form of gasoline twelve times greater than the energy used to make the fuel, a figure that even exceeds forecasts for Range Fuels' production process. However, the second advantage is perhaps even more interesting. Fermentation processes, such as those used to make corn ethanol, create large amounts of carbon dioxide. This is why there are bubbles in beer. To ensure that the global warming consequences of ethanol production are minimized, the gas must be collected and stored rather than vented to the open air. As the U.S. and other countries move to mass producing replacement fuels, this task is going to be increasingly challenging.

As importantly, a biochemical process that has carbon dioxide as one of the by-products has "wasted" some of the carbon in the feedstock. Ideally, we want all the carbon in the wood to be converted to a usable hydrocarbon fuel, not lost as carbon dioxide. A process that uses a chemical pathway that avoids an output of carbon dioxide, in this case by using bacteria that produce vinegar, has a real advantage over the conventional fermentation route. One additional implication is that this approach to manufacturing ethanol will produce very substantial volumes of fuel for each ton of feedstock. Along with low process costs and the ability to process millions of gallons a day, a high yield from the raw materials is a vital characteristic of any technology hoping to displace gasoline in the world's cars. Wood may be very cheap compared with wheat or corn—around $40 per ton, as opposed to $150 or more—but it still makes financial sense to get as much ethanol out of it as possible.

The third advantage of the ZeaChem process is that microbes eat the cellulose in the wood, and the remaining lignin is gasified into hydrogen. Virtually nothing remains as a waste product.

Although many of the start-ups making ethanol from wood claim to have this feature, ZeaChem's claim has more plausibility than most. In a world where disposing of large quantities of any waste material is getting increasingly difficult, this asset is an important part of the attraction of moving to cellulosic ethanol. We can hope with some optimism that second-generation ethanol production will be a relatively clean process, with few serious environmental impacts.

Coskata, a company backed by General Motors, is also in the leading group of cellulosic start-ups. Coskata gasifies the raw biomass, producing a mixture of carbon monoxide and hydrogen. It then passes this mixture through a stew of microbes that consume the gases and excrete ethanol. This is relatively simple technology, and the company doesn't claim to achieve the high yields of Range Fuels or ZeaChem. But its process can use a wide variety of fuels, including old tires and municipal waste. Its refineries will also likely be cheaper to build than for some of the other early-stage technologies.

Range Fuels, ZeaChem, and Coskata all propose to use thermochemical processes to make ethanol. A second possible technique is to use enzymes to breach the defenses of the tough cellulose molecule, creating much simpler sugars, such as glucose, and then use yeasts to ferment the sugars into ethanol. Industry insiders call this the "saccharification" route, referring to the intermediate step of creating simple sugars. One major barrier to commercial progress on this road is the price of enzymes, which currently add at least 25 cents to the cost of producing a gallon of fuel. At the fall 2009 oil price of $70 per barrel, the cost of the crude used to make a gallon of standard gasoline is only about $1.65, so this added cost is a major obstacle. Additionally, processes that use enzymes tend to require large amounts of heat to crack the cellulose open, a major additional cost. Once the cellulose has been turned into sugars, yeast is used to make the fuel, in a process very similar to making corn ethanol. This means that any cellulosic ethanol process that goes the saccharification route must have refinery costs at least as great as those involved in converting corn. Woody feedstocks are much cheaper than grain, so high processing costs at the refinery are not necessarily an overwhelming problem. Nevertheless, huge efforts are being devoted to getting around these disadvantages by reducing the cost of enzymes and finding innovative ways to process sugars cheaply into ethanol. Without advances in these areas, saccharification will probably be more costly than thermochemical processes.

Other routes are possible. One of the most interesting is turning cellulose directly into molecules very similar to those in gasoline, rather than making ethanol, a gasoline substitute. Such a fuel would have the advantage of being compatible with existing gasoline pumps. It could also be used neat in the car's engine. Ethanol-fueled cars, by contrast, tend to run best on a mixture containing 15 percent fossil fuel, the presence of which inevitably reduces the carbon savings.

Businesses trying to commercialize these alternative technologies are running a couple of years behind the companies seeking to crack long-chain organic molecules to make simple alcohol, but some of their ideas show great promise. LS9, a company based in San Francisco, uses proprietary genetically modified bacteria to digest fatty acids from straw and other agricultural wastes. These bacteria then excrete an oily substance very similar to standard diesel. LS9 claims it can tweak genes in the bacteria to slightly alter the fuels they produce. The company is targeting a fuel cost of about $50 a barrel, making it highly competitive with oil. It claims that its process uses very little energy for each gallon of fuel, meaning a net savings of about 85 percent of carbon emissions. It also points to another important fact: its products are far cleaner than conventional fossil fuel–derived products, meaning that carcinogens such as benzene are completely absent from the fuel and from the waste products of the production process. However, even the company says it will be at least 2011 before it will be in a position to make industrial quantities of its bacteria-produced fuels, although small quantities will be available in 2010.

Chemists can envision several other pathways by which the cellulose molecule can be converted into simple motor fuels. We will probably see two or three other processes reaching the stage of pilot plants to test whether the chemistry works in commercial volumes. Perhaps $2 billion or $3 billion of private and public capital will be ventured on experiments with cellulosic ethanol technologies. This amount may seem large, but in the context of the size of the market for gasoline in the U.S. alone—over $400 billion a year—these sums are little more than small change, which should make us optimistic. The rewards for a successful company are so enormous that capital will continue to flow into the cellulosic ethanol industry until a solution is found to the relatively simple chemical engineering challenges discussed in this chapter.

My guess is that by the end of 2011, one or more producers will be refining a cellulose-based substitute for gasoline at costs that are competitive with oil at prices as low as $40 a barrel. However, it is not just the chemical engineering that matters; equally critical is the question of whether the world has enough surplus biomass to make ethanol a serious competitor to petroleum-based fuels.

A CELLULOSE BOTTLENECK?

The crucial question for the proponents of alternative fuel sources is this: are there sufficient amounts of unused plant matter, not useful as food, to meet the gargantuan needs of the private car? Vinod Khosla is convinced there are. In a recent paper, he accepts that the land use issue is a serious challenge for ethanol. He says that we should encourage cellulose production for conversion to fuel only if little or no additional land is required so that the impact on food production is minimal. But the U.S. needs over 1 billion tons of biomass a year to replace its gasoline use, even with optimistic assumptions about the possible yields of converting cellulosic material into fuel.

The scale of the task is enormous. To meet current U.S. gasoline demand, the amount of biomass needed will almost certainly require about 200 million acres, an area larger than the farmland devoted to crops today and about the same space as occupied by U.S. national forests, or the whole of Texas. If cellulosic ethanol is used to power fuel cells providing homes and offices with

193

electricity and heat, even more land would be needed. However, the more electric cars there are on the road, the less severe the problem becomes.

The huge amounts of wood and waste needed for a gasoline substitute do not deflect Khosla's optimism. He identifies three important sources of plant matter rich in cellulose that he thinks can provide the biomass required: winter cover crops, forest wastes, and dedicated energy crops on marginal land not used for food.

Cover crops are used to maintain and improve soil structure. Planted after the main crop has been harvested, they can be left in the soil over winter. These crops are frost resistant, and the green matter can be harvested in early spring. Khosla believes that cover crops might be able to provide 20 percent of the total need for biomass. Land with cover crops can still be used to produce food in the summer, so there is no cut in food production. Unused wood from forests—wastes and trees that would otherwise have been simply left to rot—might add another 20 percent, with most of the rest coming from crops grown exclusively for their energy value. In the U.S., the most appropriate energy crop is probably switchgrass, a perennial grass that grows happily on otherwise unproductive or even degraded land. In northern Europe, the most likely candidate for this role is miscanthus, a 12-foot-high oriental grass that produces more weight of cellulosic material per acre than any other crop at temperate latitudes.

Khosla carefully lays out his view that improvements in crop yield and in forest management can produce enough cellulosic material every year. He also correctly points out that if all ethanol is made from woody wastes, the U.S. will no longer need to divert corn from the food chain as it does at the moment. This important change will increase the amount of cropland available, since a quarter or more of the best U.S. corn-producing acres are growing crops for turning into fuel. Cellulosic ethanol could therefore actually increase the amount of land available to grow food.

Nevertheless, if Khosla's faith in the potential for cellulosic ethanol is correct, a large percentage of U.S. land will still be needed to produce biomass for fuel. The incentives for land owners are probably substantial enough already. At 2009 oil prices, a biorefinery can probably afford to pay the $60 a ton that Khosla says farmers will demand for their materials and still undercut the cost of fossil fuels. Khosla is also optimistic about future improvements in biomass yield, persuasively pointing out that while yields of food crops have doubled or tripled over the past decades, largely as a result of plant breeding and better agricultural practices, virtually no attention has yet been paid to making similar improvements for the plants and trees that will be used to produce biomass for cellulosic ethanol (or indeed wood for heat and power plants). The early results from experiments into breeding faster-growing trees have produced extremely successful results. There is very good reason to believe, therefore, that today's yields will significantly improve, reducing the land area that will need to be given over to the new woody crops, perhaps by a factor of two or three.

Biomass such as dried grasses or wood chips is expensive to transport, not least because it is considerably less dense than coal or oil. For that reason, cellulosic ethanol refineries will be placed near the land that provides their raw materials, whether forests or currently unused pastureland. The need to operate large refineries to ensure that operating costs are at their minimum means that each plant might need as much as 4,000 square miles of land producing its feedstock.

This figure makes the scale of the task clear. If cellulose yields double, sufficient biomass to replace the U.S.'s gasoline demand of 160 billion gallons a year will still require almost 120 million acres, or half the land currently given over to crops. This figure is possible to achieve, but the landscape around the world would look very different in thirty years' time as traditional slow-growing trees and pastures are replaced with crops like switchgrass or miscanthus and paulownia trees in the tropics. Khosla tells us to welcome this change: it will provide communities in the developing world with a good source of income and revive many of the depressed rural areas in the U.S. and Europe. But as with the opportunity afforded by biochar (see Chapter 9), land use changes around the world are going to be enormous and potentially very unpopular. The few acres of tall miscanthus now growing in central England are widely disliked simply for being so different from the crops conventionally grown. When hundreds of square miles are given over to this tropical grass, we can expect much greater antagonism. But unless we decide to move very quickly to electric cars that can be powered from renewable energy, we will need huge acreages to be devoted to the fastest-growing energy crops, whether or not we like their appearance.

Of course, we can also hope to reduce the amount of fuel needed for each mile traveled. If engines are redesigned to run on ethanol, they will operate at higher compression ratios, and fuel economy will be better. Smaller, lighter cars will also help. These improvements require manufacturers and legislators to aggressively support new technologies for improving fuel economy. Of course, the quicker the world moves to electric cars, the smaller the need for us to grow cellulose for conversion into ethanol.

Eventually, we will probably find that batteries are a better method of propelling cars. The typical driver makes very few long-distance trips a year, and so, even if batteries continue to have limited storage capacity, the occasions on which people are going to be inconvenienced by needing to recharge en route will be limited. In the U.K., the average person only makes twenty-eight car journeys a year of greater than 25 miles, less than 7 percent of the total number of trips taken. Most people will be able to use electric cars. Commercial drivers may need to have cars that drive longer distances and so will continue to use liquid fuels, but there are relatively few of these drivers, although the distances they travel are far greater than the average.

We will also need cellulosic ethanol as an energy source for decentralized power plants, such as the fuel cells for office buildings and data centers discussed in Chapter 4. An office worker even in an energy-efficient building will typically need as much fuel for heat and power as for personal transport. Widespread use of renewable energy in fuel cells will inevitably require large acreages of land devoted to producing material rich in cellulose.

The use of land for creating cellulose, or indeed the biochar discussed in Chapter 9, is going to be part of a much larger movement to implement agricultural practices that help maintain the carbon content of the soil. Conventional agriculture, both in developed and developing countries, does not emphasize the long-term maintenance of soil quality and the retention of soil carbon. This attitude has to change, both to increase food productivity and to prevent climate change. The agricultural practices in fifty years' time will probably involve much more crop rotation (alternating different crops so as not to deplete the soil) and the mixing of different crops in the same field, combining plants grown for their energy value with those grown for food.

The perfectly understandable push to increase food yields at almost any cost over the last few decades has produced monocultures that are highly susceptible to losses from disease and from pests. So although the move to very large-scale production of energy crops for making liquid fuels will involve substantial changes in land use, the world can probably cope without reducing the amount of food produced, provided we see substantial improvement in fuel economy, a switch to electric cars, and better agricultural practices. The trickiest question is probably not whether we can grow enough biomass to fuel our cars but whether the world's agricultural land can both feed the poor and devote increasingly large amounts of primary food production to fattening animals for meat as the global population gets more prosperous. There is no easy technological cure for the impact of

197

the meat-eating habits of the rich world on the price and availability of nutrition for the poor.

CAPTURING CARBON

Clean coal, algae, and ambient scrubbers

POWER STATIONS produce a large fraction of the world's carbon dioxide emissions. In developed countries, more than a third of the total greenhouse gas output typically enters the atmosphere from the smokestacks of fossil fuel power stations. Although capturing and storing this carbon dioxide is probably the single most important thing we could do to reduce emissions, power station operators have been slow to invest in research to show how carbon capture can be carried out economically. No one doubts the technical feasibility of separating carbon dioxide from the other gases and then storing it underground. The best example of this practice is the Sleipner gas field in the Norwegian North Sea, which separates the carbon dioxide and then stores it in an aquifer. However, it is a costly and complex process that must be replicated in thousands of power stations around the world.

Mention carbon capture to an environmentalist and the reaction will usually be unfavorable. Burning fossil fuel in a power station, collecting the carbon dioxide emissions, and then pumping them underground does not seem like an ideal response to the need to reduce greenhouse gases. "It just deals with the symptoms rather than the causes" is a typical comment from climate change activists. Their view is that electricity should be generated from renewable sources and that capturing carbon dioxide from coal or gas power stations is simply a means of delaying the much-needed switch to low-carbon sources of power.

However much one might sympathize with this opinion, carbon capture is going to play a vital role in tackling climate change. World demand for electricity is increasing rapidly, and the growth of renewable energy sources is simply not keeping up with the rate of growth. In other words, the percentage of electric power coming from fossil fuel sources is actually increasing today rather than decreasing, largely as a result of Chinese industrialization and the availability of cheap coal in many parts of the world. Whether we like it or not, no successful attempt to cut global emissions can succeed without deploying equipment to capture and store the emissions from existing and future power plants. Finding the right technology for coal power stations is particularly important: a unit of electricity generated from coal produces about twice as much carbon dioxide as a natural gas power station.

Carbon capture and storage, usually known as ccs, is the subject of intense interest among coal and electricity industries around the world. But, as yet, no working power plant has installed any form of large-scale ccs. The reason is simple. Capturing the carbon dioxide, liquefying it, and then transporting it into safe long-term storage is expensive and technically difficult. A power station putting ccs equipment in place would be adding a substantial cost burden. Although the precise cost is not yet known, it is likely to work out at more than $35 for each ton of carbon dioxide, adding over 3 cents to the cost of generating a kilowatt-hour of electricity, increasing coal generation costs by 40 to 50 percent. Without a substantial and guaranteed financial incentive, no power station owner will likely voluntarily move to CCS.

Forward-looking coal-fired power station operators are almost pleading with governments to ensure high carbon taxes in order to create such an incentive. Make carbon emissions costly enough, and profit-maximizing power stations will have an incentive to install capture equipment rather than pay for their carbon dioxide pollution.

"I am a carboholic," wrote David Crane in the *Washington Post*. Crane is the head of NRG, a U.S. electricity generator with a portfolio of coal-fired stations. "If Congress puts in place a substantial carbon price," he said, "we will do what America does best; we will react to carbon dioxide price signals by innovating and commercializing technologies that avoid, prevent, and remove carbon dioxide from the atmosphere." He and other business leaders know that the technical obstacles facing power utilities are not insurmountable, but until emitting carbon becomes genuinely costly, utilities will drag their feet. The one major exception to this dilatory behavior is the Swedish company Vattenfall, which is already investing significant sums. Its hugely important pilot project is discussed later in this chapter.

We understand the individual steps needed before CCS can be commercialized. We know how to get carbon dioxide out of a mixed stream of gases, how to compress it efficiently and then transport it, even over long distances. The Sleipner carbon capture equipment has separated tens of millions of tons of carbon dioxide from natural gas. Unusually, the natural gas contains about 9 percent carbon dioxide when it comes out of the reservoir. This level needs to be reduced to little more than 2 percent before the gas is shipped onshore and sold to customers. Norwegian oil company Statoil achieves this reduction by passing the gas mixture through a liquid that absorbs the carbon dioxide but lets the natural gas bubble through. The absorbent liquid is then extracted and heated. The carbon dioxide boils off, and the liquid can be reused. About a million tons of carbon dioxide a year are collected this way and then reinjected into an adjacent aquifer. The reason it happened here first is largely because Norway already has a high carbon tax.

The Sleipner project demonstrates almost all the features that large-scale carbon capture at a power station will require. The exception is long-distance transport of the gas. However, the U.S. already has a large carbon dioxide pipeline network for moving gas. We just need power station operators to see carrying out the research necessary to combine these steps and then attaching the carbon capture equipment to working power stations as being in their long-term best interests. The best possible encouragement is a high price for carbon emissions.

CAPTURING THE CARBON

You might think that carbon capture in a coal-fired power station would be relatively simple. Perhaps the carbon dioxide is separated off as it goes up the exhaust chimney and is then pumped into a holding tank? Unfortunately it is not quite so easy. First of all, we need to understand a little bit about how coal power stations work.

Coal varies considerably in quality around the world, but the basic technology for transforming it into electric power in most power stations is fairly uniform. Coal is pulverized into a very fine powder and then burnt in a stream of air. The combustion creates heat, which then boils water and turns it into steam. This steam turns the turbines that generate electricity. Older power plants are only able to convert about a third of the heat energy of coal into electricity, but more modern power stations are designed to work at extremely high steam temperatures, which raises this efficiency to about 40 percent. Burning the coal, which is mostly carbon, produces large amounts of carbon dioxide and other waste gases. Some of the other gases are severe pollutants and are removed from the exhaust stream. The carbon dioxide is almost invariably sent up a chimney where it escapes into the atmosphere.

Many of today's coal plants are antiquated. The average U.S. plant is over thirty-five years old. They produce huge amounts of carbon dioxide compared with modern gas-fueled plants, but because coal is relatively cheap, these power stations are still economical to operate. Power station chimneys produce staggering quantities of gases. In terms of carbon dioxide alone, a single very big generating plant might produce 7 million tons each year, or nearly 1,000 tons an hour. In an old power station, this carbon dioxide will only account for around 10 to 15 percent of the total exhaust gases. The remainder—perhaps as much as 10,000 tons an hour—is mostly nitrogen, which has passed untouched through the combustion process.

The simplest way of capturing the carbon dioxide from this mixture of gases is to bubble it through a solution of ammonia salts, much as Statoil does at the Sleipner gas platform in the North Sea. The carbon dioxide reacts with the ammonia compounds, while the nitrogen floats upward. The ammonia solution containing the dissolved carbon dioxide is then extracted and put into a large tank. It is mixed with very hot steam, which heats the solution and drives off fairly pure carbon dioxide. This gas can then be compressed, liquefied, and sent to underground storage.

The major cost of this process arises from the large amount of valuable superheated steam that is needed to separate out the carbon dioxide from the ammonia compounds. This steam would have otherwise been used to drive the generating turbines, so more coal has to be burnt to replace it. One recent study showed that this "post-combustion" separation and compression would reduce the percentage of the coal's energy turned into electricity in an old power station from 34 percent to 25 percent. Therefore, the effect would be to reduce the amount of electricity generated from each ton of coal by about a quarter. The cost of the extra equipment to separate out the carbon dioxide represents a substantial further burden.

The latest generation of coal plants, of which only a few have been built around the world, use two new approaches to generate electricity. The carbon capture process is also somewhat different in each case. The first approach uses technology similar to that used in gas power plants. The powdered coal is first heated intensely in a low-oxygen environment. This gasification process splits the coal into hydrogen and carbon monoxide (the "syngas" described in Chapter 7). These combustible gases are then burnt in a gas turbine. Exhaust gases are used to raise steam, which drives a second turbine, this time a conventional steam turbine. These new plants, inelegantly known as integrated gasification combined cycle (or IGCC) power stations, are more efficient at turning coal into electricity than older coal-fired power stations but also much more expensive to build. For our purposes, the important fact about IGCC units is that the operator can capture carbon dioxide more cheaply than in older types of coal plants.

In this type of power station, the carbon capture process will involve taking the carbon monoxide gas coming out of the gasification stage, prior to any combustion, and mixing it with extremely hot steam. The water molecules in the steam split into hydrogen and oxygen. The oxygen reacts with the carbon monoxide to form carbon dioxide, which is then extracted. This process is therefore a "pre-combustion" carbon capture technology. The hydrogen from the steam is added to the hydrogen from the coal, and the gas is burnt.

The main energy loss in this process arises from the need to heat the steam that oxidizes the carbon monoxide. Producing a clean stream of carbon dioxide in this type of plant will reduce the amount of electricity generated for each ton of coal by perhaps 15 to 20 percent. The advantage of this approach is that this extra cost is slightly less than for the other carbon capture processes, though IGCC itself is a new and very expensive technology, not yet in widespread commercial use. The first plants have often disappointed their operators, and when IGCC power stations will become competitive with conventional plants remains to be seen.

This process for capturing the carbon dioxide is similar to how it might be achieved in a gas-fueled power station. BP investigated using the technique in a proposed new gas power plant in Peterhead, near Aberdeen in Scotland, but abandoned the plan when the British government announced that its early financial support for CCS would be entirely restricted to post-combustion technologies and therefore that BP would have to bear the full cost of the CCS equipment and the higher costs of operating the plant. Since more coal than gas is used to fuel the world's power stations, and the volumes of carbon dioxide from coal are much higher, the British policy may have made sense.

There's a third possible approach, usually called the "oxyfuel" process. If coal is combusted in pure oxygen, rather than air, the principal waste gas is carbon dioxide. There is no superfluous nitrogen. The power station therefore doesn't need to separate the carbon dioxide from the nitrogen after the coal has been burnt. After combustion, the only task is to extract the other pollutants and then compress and transport the exhaust gas.

This sounds like a better solution, and may well be for some types of coal. But there is an obvious downside: it takes a lot of energy to produce the pure oxygen in the first place. Air has to be chilled to $-200°C$ ($-328°F$) until it forms a liquid. The liquid air is then gradually warmed until the nitrogen boils off, leaving nearly pure liquid oxygen, which is then extracted and allowed to turn back into a gas. Although coal burns better in almost pure oxygen, the net loss of energy is still substantial—almost as much as with the post-combustion approach.

Power stations of this type are still in the early stages of commercial development. But if a high carbon tax were introduced,

205

we might see them widely rolled out. Importantly, oxyfuel equipment can also be retrofitted to coal power stations that currently burn fuel in ordinary air. We do not yet know its cost or likely impact on the amount of the coal consumed, but converting the older generation of power stations to the oxyfuel process may be a technology to watch.

The crucial point is that all of these processes for capturing the carbon dioxide from a power station require large amounts of additional energy, inevitably implying a cost penalty. Electricity produced by a plant with carbon capture is always going to be more expensive than that generated by a conventional power station. Unless legislation is introduced to mandate carbon capture, electricity generators will only switch to using CCS if the tax penalty for emitting carbon is higher than the cost of incremental energy used.

STORING THE CARBON

Compared with the challenge of capturing the carbon dioxide at the power station, processing and storing the gas is relatively simple. The carbon dioxide is compressed until it liquefies and is then sent by pipeline to where it is to be stored. Thus far, carbon dioxide has tended to be reinjected into gas and oil fields. It provides extra pressure, helping to push more of the oil and gas out of the reservoir. One of life's ironies is that carbon capture at the power station could therefore result in compensating amounts of extra fossil fuels being burnt as a result of this additional production.

In the longer run, there isn't enough space in depleted fossil fuel reservoirs to hold the carbon dioxide from electricity generation. However, carbon dioxide can also be injected deep underground into saline aquifers composed of porous rocks, as it is at Sleipner. The water in these reservoirs has too much salt for it ever to be useful for drinking or irrigation, so little is lost by storing the unwanted gas there.

Although research on the subject is not yet conclusive, these underground rock formations could probably hold hundreds of years' worth of global carbon output. Usable aquifers exist under most of northern Europe, for example. The carbon dioxide from the Sleipner gas field is injected into the Utsira aquifer, which in itself may have enough capacity to hold all of Europe's power station emissions for centuries. When injected into these reservoirs, the carbon dioxide will dissolve in the water, forming carbonic acid. In some rock types, such as basalts, this acid will then combine with minerals in the rock to form very stable carbonates, effectively locking up the carbon dioxide forever.

Many environmentalists and policy-makers worry about whether some of the carbon dioxide will eventually leak, returning to the atmosphere. The gas is buoyant and will try to escape upward in an underground reservoir of any type. Some also talk of the risk of escaped carbon dioxide concentrating at ground level and asphyxiating living creatures. Or it might collect in groundwater near the surface, acidifying the water supply. The chances of a dreadful accident of this type are probably low, but the safety of the carbon dioxide storage process will depend on the exact conditions of the reservoir and its capping rocks. So far, there has been no evidence that gas injected into the aquifer near Sleipner has bubbled back to the surface. We should be surprised if it did, because the aquifer is covered by a thick layer of impermeable rock. Nevertheless, carbon dioxide leakage remains a concern and is one of the many aspects of ccs that needs urgent and comprehensive research. The rather glib assurances from some energy companies and others that the carbon dioxide cannot escape are not reassuring enough.

THE COST OF CAPTURE

How much will ccs cost? And will this cost be greater or less than the price of carbon? Will there be a legal obligation or financial incentive for power station operators to install carbon capture on all coal-burning plants? To electricity companies around the world, these are pressing questions. When a company invests in a new power station, it needs to be comfortable that the plant will work productively for several decades. Huge amounts of money are at stake. AEP, a huge U.S. utility, intends to spend over $2.2 billion building an IGCC power station in West Virginia provided it can get approval from regulators. E.ON, the large German power generator, has outlined plans to put $1.5 billion into a new coal-fired power station in Britain.

If CCS is added to plants such as these, it will increase both the capital cost *and* the amount of increasingly expensive fuel necessary to generate a kilowatt-hour of electricity. So unless CCS is mandated by law, power generators will incorporate the equipment only if the price of carbon is high enough. The equation is simple—the carbon emitted from a power station needs to be more costly to the power station owner than the cost burden imposed by CCS. In fall 2009, a power station operator in Europe faced a carbon dioxide price of about $20 a ton in the European cap-and-trade scheme. Up to this date, electricity companies have been given free allowances to cover their needs, but that doesn't mean there's no value in sequestering a ton of carbon, thus reducing the power station's total carbon dioxide output. The company can then sell surplus permits in the European carbon marketplace, so reducing the carbon dioxide output by 1 ton will add $20 to the balance sheet.

Is $20 enough to create the right incentive for power station operators? A new IGCC plant being built today with no carbon capture will generate about 25 ounces of carbon dioxide per kilowatt-hour produced. At $20 a ton, 25 ounces cost about one and a half cents. So if the carbon price stays the same, the sensible power station owner will invest in CCS if the extra coal needed and the other costs incurred add up to less than one and a half cents per kilowatt-hour. An older plant might emit 20 percent more carbon dioxide per kilowatt-hour, meaning that it could justify installing c c s equipment even with a slightly higher cost penalty.

As the section on the Vattenfall pilot plant later in this chapter shows, a $20 European carbon price is not enough. It may need to be $45 or $60, sustained over several years, to get investors to take the risk and put in c c s equipment. The exact threshold for each power station will depend on the age of the plant, the type of coal used, and just how high a percentage of carbon dioxide is actually captured by the new equipment, but most analysts are confident that c c s makes financial sense for a large fraction of European coal-fired power stations when (and if) carbon dioxide permits trade consistently above the $45 to $60 figure. What about the U.S., which does not have a carbon tax? One suggestion from a group of energy companies, some of their major customers, and a group of environmental defense organizations was that new power stations with c c s should be paid a premium for the electricity that they produce. The first 3 gigawatts of capacity (roughly two power stations) should be rewarded with $90 a megawatt-hour. This incentive is high and would more than double the prices of electricity from these plants but would encourage risk-averse utilities to invest the capital to prove that c c s works. Later plants with c c s would receive a smaller premium, with the figure falling eventually to $30 a ton, or not much more than the current European price for carbon. This well-thought-through scheme is particularly encouraging because it shows that at least some of the generating companies that use coal for electricity production are confident that c c s can be made to work and that eventually the cost will fall to moderate levels.

Gas-fired plants can also use c c s, but the economics will be different, not least because a modern gas plant emits little more than half the carbon dioxide of an equivalent coal station. Nevertheless, even if c c s is used only for coal generation, the ultimate emissions savings from carbon capture will be enormous. Since, in many countries, coal-fired power stations represent over

209

30 percent of total generating capacity, and over 50 percent in the U.S., ccs is possibly the single most important technology opportunity described in this book.

Unfortunately, at current rates of progress, at least another decade will pass before ccs can be shown to work across the wide variety of power station ages, burning technologies, and coal types. The dilatoriness of policy-makers and much of the energy industry on this issue has been deeply shocking. The need for an urgent program of research has been obvious for five years or more, but very little has actually happened.

Twenty-five years may pass before most power stations are retrofitted with equipment to capture carbon. In other words, without a huge increase in commitment from government and power station operators, ccs is not going to be a quick solution to the climate change problem. As a result, many responsible people dismiss carbon capture, saying we should focus instead on renewables and on reducing electricity demand. This approach seems short-sighted. Coal is readily available in many parts of the world and is often the cheapest source of power station fuel. ccs is the best way we have of mitigating the impact of its use. Carbon capture techniques also have the considerable advantage of being exportable. We can only capture our own wind for electricity generation, but we can give China and India the technology to remove carbon dioxide from the growing number of coal power plants in these countries. (However, it is not entirely flippant to suggest that we might show how seriously we take climate change by sending these countries a large number of free wind turbines as an alternative today.) Of course, we cannot guarantee that these countries will use ccs, since it will always raise the cost of generating electricity, but a global carbon price should ensure power station operators everywhere will eventually find that carbon capture is in their best financial interest.

Electricity use is going to continue to rise. If the prediction in this book is right, we will eventually use electricity for much of our personal transport as well as for running appliances, lighting, and machinery. Switching from inefficient internal combustion engines to battery-powered cars might increase the total demand for electricity by at least 10 percent. We can hope that, eventually, much of the world's power will come from net-zero sources, but in the meantime, it makes good sense to accelerate the R&D into carbon capture so that increases in electricity use don't simply result in more dirty coal-fired plants. Rising awareness of climate change issues, the threat of carbon pricing, and looming lawsuits in the U.S. make power station operators in the rich world less likely to invest in coal generation than they once were. For example, many of the recent plans for new coal power plants in the U.S. have been abandoned because of pressure from worried investors. Nevertheless, since so much existing power generation is coal based, and because coal-fired plants are the easiest way of adding power to the grids of India and China, we urgently need to identify the cheapest means of capturing the carbon dioxide from this fuel.

One problem is that motivating the private sector into spending the large sums required to move ccs forward will be difficult. Canny business people know that if their company were to find an improved method of carbon separation, they would stand little chance of making money from the new technique. Any substantial improvement would be almost immediately appropriated by governments (and with very good reason). When an invention is extraordinarily valuable, it is often impossible for the inventor to protect its ownership. Paradoxically, perhaps, the very importance of ccs limits the scale of private sector research. Governments need to push carbon capture technology, perhaps *211* by direct investment or by awarding enormous prizes for specific and well-defined technical advances. Or, of course, governments could simply mandate the use of carbon capture in all coal power plants by a specified date.

The U.K. announced a carbon capture competition between power station operators. Hundreds of millions of dollars were to be made available to a company making the most impressive commitment to build a small-scale plant using ccs or to fit carbon capture onto a portion of an existing plant. It seemed like a good idea, but the government's offer of money to a single power station to build a demonstration plant has probably had the unfortunate effect of actually delaying research by several years. Rather than spend their own money, the power station owners have been waiting to see if they could get their R&D paid for. A better strategy would have been to fund a large prize that awarded several hundreds of millions of dollars to the first company that demonstrated the capture of 1 million tons of carbon dioxide. Then the power station operators would have had a real incentive to move quickly.

VATTENFALL—THE WORLDWIDE LEADER IN CARBON CAPTURE

Although most private power companies are notably uninterested in funding large ccs research projects—perhaps for the reasons offered above—the Swedish utility Vattenfall is one exception. The company is owned by the Swedish state, so it isn't obligated to maximize short-term rewards to private investors. The utility started research into viable forms of carbon capture in 2001 and began constructing its first pilot plant in 2006. This 30-megawatt power station sits next to its existing coal power station at Schwarze Pumpe in eastern Germany. It started production in late 2008, and the plant will run for ten years or more, experimenting with how best to capture carbon dioxide. Thirty megawatts is very small indeed by today's standards: E.ON's proposed new coal-fired plant at Kingsnorth on the Kent coast in southern England is over fifty times as big. But the German plant is the world's first example of carbon capture at a working power station. The cost is about $100 million, a substantial sum even for a major European utility.

Vattenfall says that oxyfuel combustion will likely provide the cheapest way of collecting the carbon dioxide at Schwarze

Pumpe. The process is a refinement of the oxyfuel carbon capture technique described above. Oxygen is separated from the nitrogen in the air onsite. Coal burns too readily if combusted in pure oxygen, so the temperature is damped down by reintroducing non-flammable carbon dioxide and water vapor from the exhaust stream. The rest of the exhaust is then cleaned of contaminants, such as sulfur compounds, and cooled. When cool, water vapor condenses, leaving almost pure carbon dioxide. Compressed to a liquid, the carbon dioxide can then be safely stored.

Schwarze Pumpe burns lignite, a form of softer coal that produces even more carbon dioxide for every unit of electricity than harder alternatives. Unusually for a power company, Vattenfall itself mines the lignite locally. The fact that Vattenfall has chosen to put its pilot carbon capture plant here is particularly appropriate. If it wants to continue to extract and burn dirty lignite from the coal fields of eastern Germany, it all too obviously needs progress in carbon capture and storage.

Vattenfall plans to store the carbon dioxide in one of the rapidly depleting fields in the Altmark gas-producing region about halfway between Berlin and Hamburg. Pumping the carbon dioxide into the gas reservoir some distance away from the wellhead will help maintain the pressure of the relatively small amounts of gas left in these fields and will increase the total volume recovered, adding just slightly to the climate change problem that CCS is meant to mitigate. The gas reservoir is about 2 miles below the surface and is overlain by many hundreds of feet of an impermeable rock, making it unlikely that significant amounts of carbon dioxide will ever escape. Nevertheless, local opposition is delaying the plans for reinjecting the CO_2 into the field, even though this technique for recovering more oil has been used safely around the world for several decades.

The relatively small amount of carbon dioxide from the pilot plant at Schwarze Pumpe will be taken in road tankers to the Altmark gas field. The plan is that seven or eight tankers will cycle between the two locations, taking a total of about

100,000 tons a year. By 2016, Vattenfall plans to construct much larger ccs-equipped lignite power stations to further demonstrate the technology. If the reservoirs of the Altmark field prove suitable, the waste gas will probably be sent by pipeline from these second-generation plants. The total capacity of the whole gas field is likely to be over 500 million tons of carbon dioxide, meaning that if everything goes well, it may be able to accommodate all the carbon dioxide produced by a full-sized power plant over the entire course of its working life. Other near-empty gas reservoirs in Germany might be able to hold the waste carbon dioxide from another four large power stations. But if carbon capture works and is used at all coal-fired plants, we will need to use deep saline aquifers, not just empty gas fields.

Understanding the lessons from the Schwarze Pumpe pilot plant will take several years. In the meantime, detailed planning will begin for the much larger demonstration plant in 2010, with the intention of producing electricity there by the end of 2016. The company estimates the cost to this point at over $1.5 billion. If all goes well, an outline design for a full-sized power station with ccs will be ready by 2020, almost twenty years after Vattenfall began its research. Two decades is a sobering length of time, particularly since most other large utilities have still barely started carbon capture feasibility studies.

Major design challenges remain, but most people in the electricity business are quietly optimistic that ccs can work. Vattenfall confidently says in public that the technical problems are all solvable. "It isn't rocket science," said a senior engineer at a recent conference. The company has also increased its own estimate of how much a power station's emissions can eventually be captured from 95 to 98 percent. Nevertheless, Vattenfall acknowledges that technical improvements will be needed to bring down the energy penalty. For example, it intends to drive water off the relatively wet lignite it uses in its German power stations. Despite the difficult future challenges, the company recently published an interview with J.P. Morgan's senior analyst covering European electricity companies. He said that on the basis of estimates provided by Vattenfall, he expected that sequestering a ton of carbon dioxide would cost as little as $45. By that reckoning, the hard-bitten board members of electricity companies would say that ccs doesn't make financial sense—yet. But Vattenfall itself has optimistically said that it eventually hopes to be able to drive the cost down to below $30 when it learns the lessons from its early plants. Others aren't quite so optimistic, and analysts talk of costs of $45 to $60 for several decades to come. Whatever the correct number turns out to be, it represents the single most important figure in the policy-making debate about how to decarbonize the world economy. If the eventual carbon price is significantly above this figure, we know that power station operators will have good financial reason to install ccs equipment and will do so voluntarily. Much below this level and we can be quite sure that they won't do it except under determined legislative attack. I think we can be certain that the quickest way to cut carbon emissions from the single largest source of emissions—the world's power stations—is to use the carrot of a high carbon price, not the stick of legislation.

Governments are now beginning to realize that carbon capture offers significant prospects for carbon reduction but that much research remains to be done. Vattenfall's commitment to the oxyfuel approach may be appropriate for lignite, but conventional post-combustion techniques may be better for the bituminous coals that are more commonly used in power stations. The correct solution may differ depending on the size of the plant, the space available, the local price of coal, and many other factors. In West Virginia, for example, American Electric Power (aep) is working with the French engineering company Alstom and the U.S. energy research company Battelle to add carbon capture to its Mountaineer coal-fired electricity plant. aep is using a patented process that absorbs the CO_2 in a low-temperature ammonia solution, heats it to drive off the gas, and then pipes the liquefied captured CO_2 a mile underground to a porous sandstone layer. Thick, impermeable shales sit above the sandstone, ensuring that CO_2 is very unlikely to escape. At present, AEP only captures a small fraction of the flue gases but, like Vattenfall, is intending eventually to equip entire plants with carbon capture. Another twenty years may pass before all these questions have been answered, but this isn't an argument for delaying research and development. When the history of the battle against climate change is written in a hundred years' time, Vattenfall and AEP's commitment to investing in carbon capture before it was commercially necessary will be seen as one of the most important steps in the move to a low-carbon economy.

OTHER WAYS OF CAPTURING CARBON

Capturing carbon dioxide using industrial equipment is not the only promising way to sequester carbon. Another approach is to feed the carbon dioxide from power stations, or simply from ambient air, to an unlikely environmental hero: algae. This idea is a form of "biofixation," just like the techniques described in the following chapters, that encourage the planet to store more carbon in its soils, plants, and trees.

The main attraction of algae—a group of several thousand water-living organisms, ranging from large seaweeds to single-cell plants—is that they are extremely efficient at breathing in carbon dioxide. Most plants use only 1 or 2 percent of the light energy they receive from the sun to productively power the photosynthesis process. Some plants, such as corn, are better than others, but all land crops are very wasteful in the way they use light. Algae, by comparison, grow faster and capture more carbon dioxide. Under controlled cultivation, the weight of algae can more than double in a day. Provided light, water, and nutrients, including carbon dioxide, are available, this exponential growth can continue forever. It is possible that using algae to capture the carbon dioxide from a power station is a cheaper way of reducing emissions than all the expensive industrial processes described so far in this chapter.

But since the algae will all die eventually and return the carbon to the air, how can these strange organisms help the climate change problem? The answer is that many types of algae have the additional advantage of turning some of the carbon from the air into usable oils. Under some conditions, up to half the weight of certain types of algae is a form of vegetable oil. After extraction, this oil can undergo simple modification and then be used as fuel in standard diesel engines. The extraordinary fecundity of algae means that they create far more usable oil than conventional plants covering a similar area. One favorite industry statistic is that an acre of algae ought to produce a hundred times as much biodiesel as an acre of soy beans.

The implications of this potential are extremely attractive. A car that runs on biodiesel made from algae will be essentially net zero from an emissions point of view. Yes, the action of burning the diesel will still result in carbon dioxide from the exhaust. But the gas will have previously been extracted from the atmosphere by the algae. In terms of the net effect on carbon dioxide levels, it would therefore be exactly equivalent to capturing the carbon dioxide from power stations and storing it underground.

In late 2007, the Anglo-Dutch oil company Shell invested in a new venture in Hawaii, where a start-up company is creating large tanks of open-air algae in coastal lagoons for eventual conversion into biodiesel. Unlike fossil diesel, this fuel contains no polluting sulfur and is harmless if spilt on the ground. No agricultural land is lost in its production, so the fuels will not reduce aggregate food production. In fact, algae may help us deal with the threat of long-term food shortages; the part of the algae that is not used to make diesel can be used as feed for animals.

Shell is backing one way of growing algae: pools in the open sea. Algae can also be grown in inland ponds or in specialized bioreactors that keep the algae inside transparent plastic tubes.

One company, Solazyme, is even intending to make batches of algae in the dark, creating growth by feeding the product with sugars rather than relying on photosynthesis. In all of these examples, the product can be harvested, dried, and then used for fuels and animal food.

Some companies look to go even further. They think that the best source of carbon dioxide for fertilizing algae is actually the unmodified exhaust gases from coal and gas power stations. Flue gases, which as we've seen contain a maximum of about 15 percent carbon dioxide, can be bubbled through water and algae. The organisms extract large amounts of the carbon dioxide to feed their growth, and very little is left to emit to the open air. Could this be a cheaper and less energy-intensive way of separating the carbon dioxide from the harmless nitrogen coming out of coal-fired power stations?

The whole idea seems almost too good to be true, and indeed, the last few years have seen many false dawns for those who believe in algae as a means of capturing carbon from power stations. The entrepreneurs working to commercialize the bio-fixation of carbon dioxide have faced setback after setback. One of the main problems has been that algae do not respond well to industrial cultivation. In large open ponds, controlling the water temperature is difficult, and undesired species of algae can take over, reducing the useful yield. The growth process in enclosed bioreactors can also be difficult to control, and one famous large-scale experiment in 2007 saw excessive growth rates in the algae physically overwhelming the apparatus installed at a large power plant.

One of the many companies trying to succeed in harnessing the power of algae is Colorado-based A2BE. The company has designed long tanks, enclosed in clear plastic, along which cylindrical rollers gently push the growing algae. One of the company's founders refers to the need "to think like algae"—in other words, to understand that this green slime is part of the natural world and

will not necessarily accommodate itself easily to artificial manufacturing processes. His business continues to have ambitious
plans to deliver huge carbon reductions. He says that an area as
small as 58,000 square miles would deliver a reduction of almost
4 billion tons of carbon dioxide a year, well over 10 percent of
today's total global emissions.

But will algae biofixation ever be successful on a large scale?
There's no doubt that we can make algae grow, that this process
absorbs carbon dioxide, and that the oils in the organism can
be extracted for fuel. What is uncertain is whether the process
can be made economically viable on a large scale. Of course, this
partly depends on the price of fossil fuels such as diesel. The U.S.
federal government terminated major investments into research
on algae decades ago because it looked as though the price of
diesel from algae would never fall much below $3 a gallon. That
price is only a little more than what U.S. consumers are paying
at the pump, and some of the scientists whose work was abruptly
stopped twenty years ago are now back in demand as consultants
to the universities and private companies furiously trying to overcome the problems they are facing with large-scale cultivation.

As with several of the technologies in this book, the case for
large-scale and sustained research around the world is overwhelming. Biodiesel from algae involves few, if any, of the
problems of biofuels made from foodstuffs. It doesn't encourage
deforestation, nor does it use a large amount of energy to grow
and then refine. Its potential production rates per acre are a
large multiple of what we can achieve with palm oil or any other
tropical plant. Perhaps instead of trying to develop algae on an
industrial scale, with huge plants covering many square miles,
we should try to farm it on a much smaller scale, using very simple equipment. This approach might mean lower yields per acre,
but it would allow farmers around the world to diversify a few of
their acres into algae for use as biofuels and as an animal food, or
even as a fertilizer for soil.

But we shouldn't give up easily on larger-scale algae plants. A quick look at the figures for the possible effect on carbon dioxide levels shows why. Ten square feet of water can grow 2 ounces or more of algae a day if fed reasonable supplies of nutrients, including carbon dioxide. That means well over 400 pounds per acre. The world uses about 80 million barrels of oil a day, and to completely replace all this crude with diesel fuel made from algae, we would have to use about 74 million acres, about 4 percent of the area of Brazil or slightly more than the size of the United Kingdom.

This challenge would be enormous, but it would be perfectly possible to achieve, should the globe's leaders decide to focus on biofixation of carbon dioxide. The reduction in greenhouse gas emissions would be equivalent to at least 25 percent of today's global total. And because most algae grow best in strong light and can cope with saline water, some farms could be placed in hot deserts with salty aquifers beneath them. The amount of water needed is not large, and it can be recycled many times. The most efficient way of fertilizing the growth is probably feeding the algae with the exhaust gases of power stations, though any source of carbon dioxide will do just as well. The other nutrients that the algae need include phosphorus, a mineral also needed to fertilize conventional farmland. Phosphorus can either be mined or, more sustainably, processed from human waste, which contains higher concentrations of phosphorus compounds than the rock extracted from most mines. In fact, human solid waste may turn out to be the best source, since mining phosphates is becoming increasingly expensive and difficult. The best possible locations for algae farms will therefore be next to power stations and close to sewage farms—land that, for obvious reasons, tends not to have high value for other uses.

It is too early to make a confident prediction but biodiesel made from crushed algae may turn out to be cheaper than ethanol made from cellulose. This would be a good outcome for the world, since growing a ton of algae will use far less land than a ton of wood or grasses. Indeed, anybody wanting to bet on which technology will win Sir Richard Branson's $25 million prize for removing a billion tons of carbon dioxide from the atmosphere might well consider a wager on sequestration by algae.

Biofixation of carbon dioxide doesn't have to use algae. Horticulture can also be good at using carbon dioxide. On England's cool and cloudy northeast coast, 24 acres of greenhouses owned by supermarket supplier John Baarda grow tomatoes all year round. The greenhouses are heated with waste heat from a nearby fertilizer plant, but the most important innovation lies in the use of the carbon dioxide that is also a waste product from the factory. Over 12,000 tons a year of high-purity gas is pumped into the greenhouses instead of being vented to the air, approximately doubling the ambient levels of carbon dioxide in the greenhouse atmosphere. The millions of tomato plants absorb the carbon dioxide through photosynthesis as part of their growth processes. This practice isn't really carbon sequestration because the carbon dioxide will return to the atmosphere when the plants die and people digest the fruit. But the 7,000 tons of tomatoes produced every year in this greenhouse complex are replacing fruit that would have been grown elsewhere, probably using much higher levels of artificial fertilizer. Since fertilizer production creates large amounts of greenhouse gases, horticultural reuse of carbon dioxide is an interesting and underexploited way of preventing emissions.

John Baarda uses waste carbon dioxide from a nearby factory, but others are focusing on taking the gas directly from the atmosphere, a technique picturesquely known as "ambient scrubbing." Carbon dioxide is only 0.04 percent of the total volume of the atmosphere, so most people think that this approach makes little sense. Surely, they say, it is easier to capture the much more concentrated carbon dioxide coming out of power stations. But Global Research Technologies (GRT) in Tucson, Arizona, believes it has found a way of cheaply and effectively capturing carbon dioxide directly from the air. The company has formulated a plastic that attracts and holds carbon dioxide molecules. When the strips of the plastic are fully loaded with carbon dioxide, they are placed in a humid atmosphere. The plastic also strongly attracts water molecules, which push the carbon dioxide away from the strips so that it can then be captured.

Although GRT was set up to focus on large-scale carbon capture and storage from the air, the company plans to demonstrate its approach and generate an initial income by producing products for the horticultural industry. One of the advocates of the technology—the eminent climatologist Wally Broecker—told me that the initial design is like a big waterwheel, half in and half out of a greenhouse and covered in strips of the plastic. As it rotates, it picks up carbon dioxide in the dry external air. The strips enter the humid air inside the greenhouse, and the water drives off the carbon dioxide, raising carbon dioxide concentrations in the greenhouse to perhaps twice the level of the air outside. The growing plants then capture the gas. Since carbon dioxide delivered in trucks to greenhouses costs over $100 a ton, the wheel will pay for itself quickly.

After the technique is proven, GRT intends to produce units the size of shipping containers that can collect at least a ton of carbon dioxide each day from the atmosphere, to be sequestered underground, just like CCS at power stations. The indicative price of these units is currently about $100,000, meaning that at the fall 2009 European price for carbon of about $18 per ton, they would take roughly fifteen years to pay back their owners. Clearly, substantial further cost reductions are necessary. The GRT machines would only take up a small fraction of 1 percent of the land area of a densely populated country, so they would use far less land than wind turbines that had a similar effect on atmospheric carbon dioxide concentrations. Additionally, there will also be a need for a pipeline network for carbon dioxide so that the captured gas can be sequestered in saline aquifers or in

222

the deep ocean, but this network is no more difficult to engineer than the reinforcement of the electricity grids required to deal with higher levels of renewable energy production.

So there is a strong case for ambient-scrubbing devices, but we would need two million of these machines just to counterbalance the emissions of a single large European country. The cost for the U.K. alone might be nearly $160 billion, or about 10 percent of one year's GDP. Is this too much? It depends how important you think averting climate change is, compared with other objectives. The London Olympics will probably cost about $24 billion when all the bills have come in, enough money to counterbalance nearly a sixth of the U.K.'s emissions for a generation if it were spent on GRT's machines instead. It is still too early to say, but ambient scrubbing may turn out to be competitive with CCS algae farming as a way of reducing concentrations of carbon dioxide in the atmosphere. Perhaps more importantly, if the world eventually panics at the sudden onset of obviously destructive climate change, building hundreds of millions of machines to actually take carbon dioxide out of the air may be the quickest way of beginning to reverse human impact on the environment.

The next two chapters look at whether soils and forests represent attractive additional ways of collecting and storing carbon. The major advantage of using the biosphere in this way is that carbon contained in trees and soils does not need to be expensively transported in pipelines and pumped into aquifers. Nor does it need millions of large devices across the globe. In addition, if we use the land to store carbon and we can find a way to reward landowners for their efforts, we will help to engineer greater involvement of developing countries in the global battle against climate change. Paying many of the world's poorest people $100 a year in return for helping us store carbon may be cheaper than any form of industrial carbon capture. We will also be compensating these people for inflicting climate change upon them. As I try to show, carbon capture in soils and through prevented deforestation will also help improve agricultural productivity. Equality demands that where possible, we should prefer to use carbon dioxide reduction techniques that improve the living standards of the poorest people in the world, particularly those already suffering from diminishing rainfall and lower grain yields.

This is not an argument against capturing carbon emissions from power stations or using ambient scrubbing technologies such as GRTs. We need the widest possible portfolio of techniques for holding down emissions and then reducing carbon dioxide in the atmosphere. ccs at coal and gas power stations is a vital ingredient in any carbon-reduction plan.

BIOCHAR

Sequestering carbon as charcoal

DEEP IN the Amazon jungle are unusual patches of land where soil is darker and richer than in the rest of the region. These areas are highly fertile and also contain large quantities of carbon—carbon that has been drawn out of the atmosphere and safely locked away for hundreds or even thousands of years. These so-called terra preta ("dark soil" in Portuguese) hold the key to one of the most exciting ideas in the fight against climate change. It's not about high-tech panels, turbines, or vehicles. It's about rethinking the way the world uses a simple and familiar substance: charcoal.

The last chapter was largely about carbon capture using industrial processes. This chapter and its sequel deal with ways in which we can permanently store increased amounts of carbon dioxide in the world's soils and vegetation. Through the photosynthesis process, plants and trees naturally take in carbon dioxide as they grow. Carbon from the air is used to make the complex sugar molecules that serve as the physical structure of these living organisms. When they die, the carbon absorbed by the trees and plants generally returns to the atmosphere, either through burning or through gradual rotting. This process is often called "the carbon cycle" and has been going on since the beginning of life on earth.

One way to beneficially disrupt the cycle is to partly combust wood to make charcoal, an almost pure and extremely stable form of carbon. This chemical stability means that unburnt charcoal sequesters carbon for centuries, even if it is simply mixed in with the soil. So, if we make charcoal from wood and then dig it into the soil, we are sequestering carbon from the atmosphere just as much as if we were capturing it at a power station.

There are far too few quirky and unexpected ideas in the climate change field. Taking carbon dioxide from power stations to feed oily algae to make biodiesel is one such strikingly neat suggestion. Making charcoal, or "biochar" as it is often called, is another. Previously skeptical scientists who have examined the impact of charcoal on the carbon cycle conclude that it does seem to permanently remove carbon dioxide from the air. Crucially, mixing charcoal in with soil has a very beneficial side-effect: it can significantly improve agricultural yields, particularly for topsoils deficient in carbon. As the evidence of biochar's effectiveness increases, research interest is growing around the world. This chapter looks at the impact of biochar added to arable soils, primarily in the tropics. The following chapter examines other ways of adding carbon to soils and forests.

Unlike many of the technologies discussed in this book, biochar doesn't necessarily require expensive equipment or highly skilled people. Capturing and processing the carbon dioxide from power station chimneys is technologically complex and expensive. By contrast, making charcoal is easy and can be profitably done by poor farmers in the tropics.

It was the terra preta area of the Amazon that gave rise to the idea of adding charcoal to the world's soils to reduce carbon dioxide levels. Found spread all over the Amazon basin, these fertile patches are sometimes less than a few acres in extent, but in places they cover several square miles. Perhaps as much as 10 percent of the region has soils with substantial amounts of added carbon. Terra preta exists in a wide variety of different soil types, but they all share the characteristic high levels of charcoal residues. Research scientist Bruno Glaser from the University of Bayreuth in Germany says that an acre of terra preta soil three feet deep typically holds 100 tons of carbon, compared with 40 tons in adjacent soils that have not been improved with charcoal.

The high fertility of the dark soils of the Amazon basin has been known for decades. Writers were commenting on it approvingly in the nineteenth century. Scientific attention first arose when the Dutch soil scientist Wim Sombroek published a highly influential book in 1966 on the human-made soils of Amazonia. He showed convincingly that the impressive soil productivity was a consequence of the high carbon content. Sombroek had grown up on a farm in the Netherlands, which had "plaggen" soils—rich, deep, and highly fertile. These soils had been created over centuries by local farmers adding thin, carbon-rich turfs covered in cow slurry to the existing surface. He found that the Amazonian farmers had similarly adjusted the soils to provide long-lasting improvements to their fertility.

With growing enthusiasm over the last few years, other researchers have noted that charcoal can improve agricultural productivity in many different types of soil around the world, doubling or even tripling yields in some circumstances and climatic conditions. We certainly do not completely understand the process by which biochar aids fertility. More research is needed, but evidence is mounting that biochar's highly porous structure helps retain valuable nutrients and provides a protective structure that encourages beneficial microfungi to grow. Look closely at a piece of barbecue charcoal, and you'll see thousands of tiny holes that were the cell walls of the original wood. This sponge-like structure offers a huge surface area that, it seems, helps make nutrients available and provides useful support to which benefi-cial organisms can cling. Increasingly, we also understand that in the dry tropics, biochar helps the soil retain water and therefore helps crops grow, particularly in times of drought.

The structure of ground-up char may also help prevent valu-able plant foods from leaching into streams and rivers. Biochar that has been laced with potassium, phosphorus, or ammonia before being added to the soil appears to achieve even better results than simple charcoal. In some ways we don't yet fully understand, the charcoal acts as a catalyst, making nutrients available to plants without being affected itself. Simply because it improves soil fertility, adding biochar to a large percentage of the world's soils may make good sense, even before we consider the potential impact on atmospheric carbon dioxide. Laurens Rade-makers, a Belgian social scientist with an interest in economic development in the tropics, set up an important trial in the West African country of Cameroon, where weather and soil conditions allow two crops of corn to be grown each year. For each of the two 2009 corn-growing seasons, he persuaded many groups of subsistence farmers to apply large amounts of biochar to some of their plots and to compare the results with equivalent plots that hadn't been dosed. The biochar was made from the agricul-tural wastes of a previous crop. The waste would usually have been disposed of in open fires, so making biochar did not divert organic matter from other uses. So far, the results of the trial have demonstrated the impressive ability of biochar to reinvigorate degraded and vulnerable croplands. Those plots that received the most biochar had approximately doubled food yields. This means that biochar had as much of an effect as adding large amounts of expensive fertilizer. When fertilizer (organic and mineral) was also added, the yields were even better.

Rademakers is far too good a scientist to claim that one set of trials, albeit a large and well designed one, is enough to prove that biochar works, but the improvement in yields certainly supports a hypothesis that it has a beneficial effect and provides a use for wastes that otherwise would simply be burnt. However, big problems remain. The first is determining how to move from a handful of small-scale plants making biochar to exploiting the idea all around the world and sequestering billions of tons of carbon each year. The second is to find out why biochar doesn't invariably add to soil productivity. Occasionally, fertility may even decrease. No one has yet worked out why. We will need much more research on what plant materials should be used to make the biochar, the right temperature for the charcoal kiln, and what other fertilizers should be added to the soil at the same time, perhaps already bound to the biochar. But these are questions that science should be able to solve without too much difficulty. Biochar stands as a good prospect of being one of the simplest, cheapest, and most effective ways of capturing carbon dioxide from the atmosphere and storing it safely.

BIOCHAR AT THE SMALLEST SCALE

Irishman Rob Flanagan is one of a growing network of practical idealists around the world working on biochar. His life changed when he saw a 2002 science program on BBC television about the terra preta soils of the Amazon. Excited by the visible evidence that biochar could help apparently infertile soils support productive agriculture, Flanagan went to work for EPRIDA, a pioneering biochar company just outside Atlanta, Georgia. After a couple of years working there, learning how to make charcoal, he carried out his research in tropical China and in Indonesia.

Biochar can be made in tiny kilns that double as highly efficient cooking stoves, or it can be made in huge chambers that produce hundreds of tons a day. Flanagan's interest is in biochar at the smallest scale. His mission is to design a simple, cheap, and reliable domestic cooking stove that uses locally available materials for its fuel and, as a by-product, gives the homeowner some charcoal to feed the soil. Some—perhaps a large part—of the world's deforestation is being caused by families seeking wood for cooking. Flanagan wants tropical households to have fuel-efficient stoves that could reduce the amount of wood needed, sequester carbon, *and* boost soil fertility.

The smoky and inefficient open stoves in use in much of the developing world are highly wasteful of wood, increasing the amount that needs to be cut down and raising the amount of carbon dioxide released into the atmosphere, as well as reducing air quality inside and outside homes. About 1.4 billion tons of wood and other organic matter are used for cooking fuel each year. If rolled out universally, efficient domestic biochar stoves, such as those Rob Flanagan is developing, might reduce this figure by a half or more. This would prevent the need to cut down virgin forest, potentially slashing global carbon emissions by 10 percent.

Importantly, Rob Flanagan's stoves can also burn agricultural wastes, such as rice husks, which would otherwise be unused. His stoves leave a good percentage—perhaps 15 percent by weight—of the original fuel as charcoal in the bottom of the stove. This charcoal could be burnt as fuel, but Flanagan believes that the best use is as a high-quality soil conditioner and fertilizer.

In one of his recent experiments, Flanagan compared the germination and growth rates of a fast-growing native Chinese tropical tree in seed trays with and without biochar. The differences were dramatic. The soil fertilized with small quantities of biochar produced much more vigorous plants. The leaves were a healthy dark green. The roots were stronger, too, and it seems likely that plants prepared this way will grow much more quickly when transplanted to their final growing location.

To Rob Flanagan, these results suggest a virtuous circle. The charcoal in the soil helps new trees to grow rapidly, which increases the amount of available wood fuel and, therefore, the volume of future charcoal. He wants to see biochar stoves in every agricultural village in Asia, reducing the amount of wood used and decreasing the amounts of money poor families spend on fuel. Authorities in regions under threat of desertification are planting billions of new trees, but some Chinese regions are still losing forest cover. This loss increases the amount of carbon dioxide going into the atmosphere and affects the local climate. The absence of respiration (the return of water vapor from trees to the atmosphere) increases the threat of drought.

Equally importantly, Flanagan's stoves burn extremely cleanly, improving the air quality in homes. The World Health Organization reports that a million and a half people die every year from the effects of indoor air pollution, which is mostly caused by smoke from open fires in poorer communities.

Like the other experimental scientists working in the field, Flanagan isn't sure why biochar adds to the soil's fertility. When I chatted to him by e-mail, he described this puzzle as the "million-dollar question." The fertilization effects of charcoal are clearest in the tropics, but many researchers are now seeing similar improvements in the soils of temperate lands. Flanagan has seen extremely good results in New Zealand, for example. But he admits that much research will be needed to work out the optimal temperature for his stoves and what types of woody fuels make the best charcoal for the soil.

It is still a frustrating time for the dedicated biochar researchers working in the tropics. They all strongly suspect that biochar has a very important role in mitigating climate change, as well as in improving living standards in poor communities. Research funding has been difficult to secure, however, and people like Rob Flanagan are working on isolated experiments that should be happening everywhere around the globe. Findings are shared via some very active e-mail discussion groups, while short videos giving the results of the latest experiments are uploaded to YouTube. But despite the remarkable effect of biochar on some plants and the huge potential for carbon sequestration, he still finds initial skepticism everywhere. "It all just sounds too easy," he says. "No wonder no one gets it."

Flanagan is trying to develop a very low-cost stove that will be easy to use and that local people will want to use for their cooking. As many other attempts to improve cooking practice around the world have shown, even the best stoves require some skill to operate. The women who do most of the world's cooking will need to be convinced that Flanagan's revolutionary stove will work. There's also the question of price. When built in large volumes, the stoves should cost substantially less than $40, but this sum is still enormous for the very poorest people to afford. Nevertheless, a stove should repay itself quickly by substantially reducing the amount of fuel that the household has to buy or collect. We can hope that microfinance banks, such as Grameen in Bangladesh, will be able to lend money for biochar stoves because of the long-term savings to the household budget. Similarly, the growing number of institutions offering voluntary carbon offsets for people in the rich world might choose to subsidize the sales of Flanagan's stoves in the poorest countries. There are fewer projects where the carbon savings are more obvious.

But getting interest from the international bodies that concern themselves with climate change is difficult. Unlike some of the technologies covered in this book, the opportunities for big international companies to build large businesses based on sophisticated technology are simply not present. Large pools of U.S. venture capital are flowing freely into solar photovoltaic technology and second-generation ethanol manufacture. But biochar, an approach to carbon sequestration that is potentially also hugely beneficial to tropical agriculture and the deforestation problem, is virtually ignored. With small biochar stoves, the low income of most of the likely beneficiaries holds back research, development, manufacture, and marketing.

SCALING UP

Rob Flanagan's focus is small-scale biochar production in poor rural communities. BEST Energies in New South Wales, Australia, operates at the other end of the spectrum. BEST has spent the last decade developing a technology that can be scaled up to partly combust almost 96 tons of dry biomass each day, generating perhaps 30 or 40 tons of biochar. The biomass doesn't have to be wood. It can be agricultural waste or even municipal garbage.

The biochar from BEST's pilot plant is added to Australian soils, which have some of the lowest carbon levels in the world, and the results have been spectacular. Lukas Van Zwieten, a scientist working for the New South Wales government, found that adding 4 tons of biochar per acre tripled the mass of wheat crops and doubled that of soybeans. Even if only part of this improvement is seen in other experiments, biochar would revolutionize the food-production potential of carbon-starved soils. The recent Cameroon results from Laurence Rademakers have yielded similar improvements.

When woody wastes are heated to high temperatures, the material gives off gases and material that will condense back into liquids. These gases and liquids can be burnt directly as fuel. The BEST biochar plant will also produce significant amounts of useful heat from combusting the by-products in this way. Hot gases can be fed into a turbine that generates electricity. As coal and gas get more expensive, using biomass to make electricity is looking more and more attractive. Chapter 4 described how a similar technology that completely gasifies beetle-riddled wood is being used as a source of heat and power in western Canada. BEST Energies' process keeps some portion of the wood as charcoal.

EPRIDA, the U.S. business that gave Rob Flanagan his first direct experience with biochar, has similar ambitions to those of BEST. It wants to build large-scale processing units that use woody wastes to create electricity and biochar. The difference is that EPRIDA believes it can make the biochar even more productive

233

by adding fertilizers into the char as it is produced. The company's process takes the hydrogen that comes out of the wood and turns it into a slow-release, ammonia-based fertilizer bound into the pores of the biochar. EPRIDA says that this material encourages the growth of a particular type of fungus that lives on the roots of most plants. These fungi feed using extremely thin hair-like tubes that gradually reach inside the pores of the biochar. The tiny tubes exude a protective glue-like substance that also binds together tiny bits of dead organic matter. This glue is very chemically stable and helps maintain the soil structure and its overall carbon content in addition to boosting plant growth.

How much difference could plants such as those BEST and EPRIDA are proposing make to carbon emission? The answer partly depends, of course, on whether biochar adds to the pressure to cut down the world's forests. Sequestering carbon in the soil, however beneficial for agriculture, is not going to reduce atmospheric carbon if the charcoal makers cut down virgin forest to make the biochar. That caveat aside, biochar plants could make an enormous difference. For the sake of illustration, let's say that Germany, Europe's largest emitter, decided to focus on biochar as a way to take its emissions to zero. Plants such as the ones offered by BEST Energies can produce 30 to 40 tons of biochar per day, so we would need about twenty thousand machines to cut emissions by 250 million tons of carbon a year—more than enough to wipe out the country's total greenhouse emissions.

If we created the right economic incentives, installing this number over a five-year period at a rate of four thousand plants a year would be perfectly conceivable. For comparison, there are also about twenty thousand wind turbines in Germany, most of which have been erected in the last few years. The crucial signal that entrepreneurs need in Europe or in a poor country reliant on agriculture is a high and stable price for sequestering carbon. While many countries now reward companies in the form of grants and subsidies for producing low-carbon electricity, no government pays money to businesses putting carbon back into the soil, an equivalent activity in terms of its effect on carbon dioxide levels in the atmosphere. As a result, commercial biochar companies are currently struggling to get the early commercial orders that will prove their technology works outside research laboratories.

LOCAL EXPERTISE

Rob Flanagan's domestic stoves and BEST's large-scale plants are at the extreme ends of biochar production techniques. A middle way is to use existing local expertise in charcoal making. The only difference from what already happens today is that more would be produced, and, instead of all the charcoal being used for fuel, some part would be diverted and dug into the soil.

Charcoal making is carried out in countless different ways around the world, but all the methods have something in common. They restrict the flow of air into a pile of burning wood or plant matter such as coconut husks. Doing so keeps the temperature down, and the lack of oxygen stops much of the carbon from burning. At the relatively low temperature in charcoal kilns, volatile gases and liquids are driven from the woody material, and these can be collected or burnt either for heat for cooking (as in the case of Rob Flanagan's stoves) or electricity generation (in industrial-scale plants). The liquid driven off, often called "wood vinegar," is sometimes separated and used for a variety of purposes, such as an insecticide or even as a health drink in Japan.

Charcoal is what remains when all the volatile matter has been driven off. Because it is almost pure carbon, charcoal itself is an excellent fuel, burning cleanly and evenly and at a high temperature.

235

Considerable skill is involved in making charcoal. Throughout history, communities have tended to employ specialist charcoal makers who earned their living making fuel for their neighbors.

The charcoal maker needs to know how wet or dry the wood should be, how to limit the flow of air into the wood and when to open up the burning pile to collect the valuable charcoal.

Some cultures make their charcoal by building a pile of wood and then covering it with dense and non-flammable matter to reduce the air flow. In many places in northern Europe, for example, charcoal is made by covering a carefully made log structure with small branches before completing the pile by covering it with earth to stop too much air from getting to the wood. Other societies use purpose-built buildings, while some make their charcoal in pits in the ground. The woody material is usually set alight by inserting a flame into a chamber in the middle of the pile. The pile may take days to turn into charcoal, and during this time it must be carefully watched to ensure that it does not combust too freely. When the charcoal is ready, the pile is broken open, and the hot charcoal is dampened with water to stop it from burning further.

When farmers and small businesses make charcoal today, they are not doing so to dig it into the soil. They produce charcoal because of its value as a fuel. In the developed world, charcoal is mainly associated with summer barbecues, but in less well-off countries, it is valued as a light and clean indoor fuel. In some regions, it is the most important source of energy for cooking. Made in the countryside, charcoal is easy to handle and, being lightweight, easy to transport to nearby cities. In historical times, it was also widely used to smelt ores for metal. For example, charcoal will burn at a high enough temperature to create liquid iron from iron ores. No doubt, in some parts of the world it is still used for this purpose.

So charcoal making is established and well understood across the globe. If we are to use biochar to help mitigate climate change, we need to massively expand the amount of charcoal being produced in poorer rural areas. But because the skills already exist in such communities, this is not an impossible dream. The crucial issue we face is finding a mechanism that rewards village-level enterprises in developing countries for producing charcoal and adding it to soils.

By any standard, the cost of storing carbon in the form of biochar is extremely low. In poor countries, many people earn very little money for their work—perhaps just a dollar a day. Charcoal makers producing biochar in larger kilns in an agricultural community might be able to make several hundred tons every year. If they were paid for their charcoal even at the current European market price for carbon dioxide, they would earn wages that they could never hope to make in other occupations. The equipment that they need is simple and available locally, and so there are no obvious obstacles to building large charcoal industries even in very poor countries.

Importantly, and perhaps surprisingly, verifying that carbon has actually been stored in soils is not difficult. Soil carbon levels can be measured simply and accurately with cheap equipment. New techniques even allow remote sensing of carbon levels using satellites. Of course, taking such measurements in millions of fields around the world each year, and remunerating the relevant farmers, will present enormous challenges. Nevertheless, the scope for carbon sequestration using biochar made in poor, rural areas is huge. If organized on a sufficient scale, it may be no more expensive than the complex engineering required for carbon capture at coal-burning power plants.

THE IMPORTANCE OF SOIL CARBON

We generally focus too much on carbon dioxide in the atmosphere. The world's soils hold twice as much carbon as does atmosphere and about 1 trillion tons more than the world's plants do. The thin layer of soil that covers much of the earth's surface but is rarely much more than 3 feet thick, produces a very large fraction of our food and sustains life as we know it.

The location of the world's carbon is not static, of course. Carbon in its various forms moves from soil to air to plant and in the reverse direction all the time. The amount of carbon—usually in the form of carbon dioxide—flowing in and out of soils in the natural cycle is perhaps ten or twenty times the volume put in the atmosphere by the burning of fossil fuels. In other words, we don't have to disrupt this cycle by a large percentage in order to achieve real reductions to the amount of carbon dioxide ending up in the atmosphere.

One further comparison should strike us forcefully. The weight of carbon in the soils is about 1.6 trillion tons. Human actions produce about 8 billion tons of carbon in greenhouse gases every year, of which about half is added to the atmosphere (causing climate change), with the rest absorbed by oceans, plants, and soil. As the planet warms, one side-effect could be a small annual reduction in the soil's ability to retain carbon. As temperatures rise, many chemical reactions in the soil will speed up, meaning that carbon dioxide, an important end-product of many of these reactions, will tend to escape more quickly back into the atmosphere. This is one of the many "tipping points" that threaten to accelerate climate change.

Worryingly, an increased flow of carbon dioxide from soil to air may be happening already. One recent paper in the science journal *Nature* suggested that the U.K.'s soils are losing carbon at the rate of 0.6 percent a year. The loss in the most carbon-rich British soils is over twice this rate. Whether the same is happening across the rest of the world isn't yet clear, but as temperatures increase and rainfall becomes more erratic in many areas, there are good reasons to assume that there will be a net shift of carbon from soil to air.

A 1 percent loss of the carbon in the soil across the globe would approximately triple humankind's total emissions for a year. We know so little about the impact of changing temperature and rainfall on the world's soil that this simple comparison should extremely concern us. By burning huge amounts of carbon-based fuels and raising the equilibrium temperature of the atmosphere, we are, as is so often said, conducting a gigantic and dangerous experiment on the planet.

Modern agricultural techniques are another reason soil carbon levels may be falling. Plowing, overuse, and erosion by wind and water all tend to reduce the carbon content of the land. One respected researcher says that some agricultural soils have lost 50 to 70 percent of their embedded carbon content since the Industrial Revolution. As more of the world's land is pressed into intensive agricultural use, soil degradation is probably becoming an important source of carbon emissions into the air.

A third cause of the loss of soil carbon is deforestation. When an acre of trees is cut down, perhaps because a farmer wants to convert the land to agriculture to feed the world's growing population, the carbon in the wood is lost, usually in a fire used to clear the land for cultivation. Flushed out by rainfall, the newly exposed soils also rapidly lose much of their embedded carbon.

The last half-century has seen truly remarkable increases in agricultural productivity. Average grain yields have almost tripled. Despite the rapid growth of the world's population, a smaller percentage of the world goes to bed hungry than ever before. But even to some optimists, this improvement shows disturbing signs of tapering off, with the dramatic recent food price increases being a sign of things to come. The supply of new land suitable for arable farming looks surprisingly limited. The Malthusian trap—food production growing slower than population—is a specter that many commentators laughed at a few years ago but that now frightens increasing numbers of policy-makers. One respected U.S. government forecast shows the growth rate of cereal production dipping below the rate of increase in global population within a few years. Rising meat consumption will compound this effect as more and more grain is diverted to feed cattle and pigs.

So if biochar did nothing else but add to the productivity of our soils, help stabilize soil carbon, and reduce the pressure to cut down forests, we would have good reason to encourage its manufacture. But done on a sufficiently large scale, it could also make a real difference to the amount of carbon dioxide in the atmosphere.

BIOCHAR'S CARBON IMPACT

At the moment, carbon dioxide levels in the atmosphere are rising by about 2 parts per million (ppm) every year. This rate is increasing, partly because fossil fuel use is growing and partly because the traditional stores of carbon—soils, vegetation, and the oceans—seem to be becoming less effective at soaking up a large slice of human-made emissions. Evidence suggests that a slightly larger fraction of the carbon dioxide generated by our actions each year is ending up in the atmosphere. In view of this development, what should our target for biochar be? Perhaps optimistically, we might look for it to reduce atmospheric concentrations of carbon dioxide by 0.2 ppm each year, about 10 percent of the current net increase. That would be equivalent to eliminating almost half the total emissions of the U.S. or of China.

To achieve a reduction of this scale, we'd need to create about 400 million extra tons of charcoal and sequester it in soils. Averaged across all types of organic material—wood, plants, shrubs, and grasses—biomass is about 50 percent carbon. So if we could convert all of that carbon into biochar, we'd need around 800 million tons of source material. In reality, the process would probably be only around 50 percent efficient, so we'd actually need to process about 1.6 billion tons, which is over 1 percent of the world's total biomass growth each year and is equivalent to a little less than a fifth of the biological material we produce on the world's croplands. Processing such a huge amount of vegetation would be a very challenging task.

240

Where would 1.6 billion tons of biomass come from? Agricultural wastes could account for part of it. Of the 9 billion tons of material produced on agricultural lands, probably less than 5 billion tons is actually food. The rest is straw, husks, leaves, and other currently unusable matter. We could use a substantial part of this waste for making char—though the benefit would be tempered by the fact that some agricultural wastes are already plowed back into the field, resulting in some increase in soil productivity and carbon levels.

Furthermore, we will need to use the biomass from the world's lands to make the cellulosic ethanol discussed in Chapter 7 and for burning in the district heating plants referred to in Chapter 4. At first sight, biochar seems to be one of a range of substantial extra demands that we will need to place on the world's agricultural, forest, and savanna lands.

This interpretation may be too pessimistic. Some of the biological material we need for biochar can come from converting wooded areas containing slow-growing trees to what will be, in effect, charcoal factories using grasses, trees, and shrubs that develop much more rapidly. The world's surface has about 39 million square miles of vegetated land, including forests, savanna, and cropland. At existing typical growth rates, to grow all the 1.6 billion tons of biomass we need might require us to convert up to 2 percent of this area, or about 490 million acres, to biochar production. This is an area four times the size of Texas.

If we got our char half from forests and half from agricultural wastes, we would still need more than 1 percent of all the world's green lands to become charcoal factories. Achieving this goal sounds like an enormous task that is well beyond our capacity to organize. But if we doubled the typical weight produced by an acre of forest by switching to faster-growing species, this task becomes much more manageable. This idea has been greeted with horror by some green commentators who envision biochar production becoming a repeat of the disaster of first-generation

241

biofuels. The proponents of biochar need to address this point—
if biochar becomes a valuable commodity, how do we stop local
entrepreneurs and large companies from cutting down forests in
order to make it? I think there are three potential responses.

First, a sustained international policy of biochar manufacture
will improve crop yields, particularly on degraded arable soils. So
biochar will help the world feed its growing number of people,
and it may make good financial sense to use charcoal to improve
soil fertility rather than burn it for fuel. Certainly, the 2009
Cameroon results of Laurens Rademakers support the idea that
biochar is so worthwhile for agricultural productivity that it will
reduce the pressure on the world's forests rather than increase it.
If we can make more food on existing arable lands, we don't need
to cut down forests.

Second, biochar will largely be made in poorer tropical
nations where wage levels are low but biomass growth rates are
typically high. It will be considerably less expensive to sequester
carbon using tropical biochar than it would be in developed, temperate countries. In fact, biochar could be a profoundly effective
form of job creation in the poorest countries of the world.

Third, the processes of making biochar also produce useful heat, which can be used for domestic cooking if operated on
a small scale. In larger biochar plants, the gases can be burnt to
provide electricity through a simple turbine. Some years will pass
before we find out whether solar or wind technologies are better
for supplying electricity in less prosperous parts of the world, but
a biochar kiln that burns wood gases in a turbine may be a cheap
way of creating a reliable electricity supply in a remote country.
Another advantage is that it can easily generate electricity at
night or when the wind isn't blowing.

Some readers will still be aghast at any suggestion that we
convert existing forests to plantations for biochar raw material.
The concern is easy to understand: anybody worried about global
warming should be eager to maintain all the woodland we have

today. But many types of tree are slow growing. If we replace these species with rapidly growing trees or grasses suitable for the region, we will eventually increase the uptake of carbon dioxide from the air. In a temperate climate with reasonable rainfall, an acre of broad-leaved trees might increase in weight by 4 or 5 tons a year. Replacing these trees with fast-growing willow or miscanthus can triple or quadruple this weight gain, thereby massively increasing the potential for biochar carbon capture.

Of course, if we created a huge worldwide biochar industry, we would need to put in place stringent measures to protect biodiversity. In the drive to maximize biomass production for biochar and ethanol manufacture, we run the risk of creating areas of dangerous monocultures. One single type of tree might be planted on every spare acre of land for hundreds of square miles. This practice would induce a catastrophic loss of diversity of animal and insect species in some parts of the world. It might also increase the possibility of disease and pest infestations that kill trees or reduce their growth. So the world will need to control the planting of fast-growing trees to ensure sufficient variety of species. Once again, no one can pretend that this task will be easy. We need economists to tell us how to create the incentive for landowners to grow as much biomass as possible while protecting soil and biodiversity and minimizing the risk of widespread disease and infestation.

Some soils treated with biochar have another impressive characteristic not mentioned so far: they seem to give off far lower volumes of nitrous oxide. Nitrous oxide is a far worse global warming gas than carbon dioxide, and, as we saw in the chapter on cellulosic ethanol, agricultural land is a prime source. In fact, for many crops grown in temperate regions, nitrous oxide emissions may be more significant than the carbon dioxide produced by food growth and processing.

This benefit is likely to be seen mainly in cooler countries and places where most artificial fertilizer is applied, though in one trial in tropical Colombia, researchers found that adding biochar to grasslands reduced nitrous oxide emissions by 80 percent. Output of methane, also a serious global warming gas, was cut almost to zero. Even if the addition of biochar is found to make little difference to agricultural productivity in soils that are already fertile, the impact on nitrous oxide emissions may be substantial. Although data on agricultural nitrous oxide emissions is poor, farming is becoming an important source of this greenhouse gas in some countries. Introducing biochar into the soil may be a cost-effective way of reducing aggregate emissions.

Scientific knowledge of the best ways to use biochar is still developing. One key researcher in the field is Johannes Lehmann, based at Cornell University in upstate New York. Lehmann and a former graduate student of his, Christoph Steiner, think that the most effective way to use biochar to combat climate change is to work to replace traditional "slash and burn" techniques of subsistence tropical agriculture with what the researchers call "slash and char." They point out that in many developing countries, farmers move from area to area, cutting down and burning the forest as they go. The cleared land delivers reasonable yields for a couple of years but then rapidly declines in fertility and becomes unusable for a decade or more. If, instead, the farmer cut down the trees but then made biochar from them in a low-oxygen kiln, the charcoal could be added to the soil and could keep it fertile for much longer.

On the best terra preta soils—which probably took several decades to create, many centuries ago—the fallow period is as little as six months after one crop has been harvested. Properly looked after, these soils never become unusable or have to be left for long periods of unproductive fallow. The slash and char technique could make a huge difference both to agricultural yields on weathered tropical soils and to the area of forest that has to be chopped down each year. The key task is to persuade millions of small-scale farmers across the tropics that slash and char is a

244

better technique than their current methods. Once again, this task may seem intimidating, requiring huge amounts of education in remote communities. But the impact on poverty and rural deprivation could repay the effort, even before measuring the global warming benefits.

Johannes Lehmann is one of the great optimists of this story, offering a sense of the enormous potential of biochar. He says that by the end of this century, we could capture 9.5 billion tons of carbon each year simply by adopting biochar manufacture on a large scale in tropical agricultural systems. This figure is striking; if we achieved this level of carbon capture today, atmospheric concentrations of carbon dioxide would be falling.

KICK-STARTING THE BIOCHAR REVOLUTION

What do we need to do to get meaningful amounts of biochar into the world's soils? In countries where we can install substantial numbers of large-scale plants, such as those that BEST Energies or EPRIDA will produce, all we probably need is for governments to acknowledge that biochar should be included in carbon trading schemes. During 2009, policy-makers certainly became very interested in this option. A ton of carbon stored in the soil has the same effect as capturing the same amount from a power station. So the operators of biochar plants need to be able to sell their carbon sequestration for the standard market rate.

At the fall 2009 price of carbon dioxide in the European Commission's trading scheme, a ton of carbon was worth almost $80. In richer countries with good soils, this price might already be enough to make biochar commercially viable. Of course, the charcoal plants would additionally be able to sell their electricity and the biochar to farmers trying to improve soil productivity. If biochar also helps reduce nitrous oxide emissions from the fertilizers added to the soil, the financial advantages would be even clearer. Building large-scale biochar plants in poorer countries

245

may or may not make sense. The capital costs of the equipment will be high, and it may be cheaper to make biochar using large amounts of labor to construct charcoal ricks.

In less prosperous countries, the fertility effect of adding biochar to the soil may be so substantial that it makes good sense even if the farmer is not remunerated for storing carbon. But in many parts of the world, people live very close to the margins of survival. Rather than use the charcoal as a soil improver for next year's crop, households may decide to burn the charcoal instead, especially in areas short of firewood. Three things must happen to make farmers in poor countries want to put charcoal on their fields. They need to be able to buy low-cost, fuel-efficient stoves on financial terms they can comfortably manage. Second, particularly for those on the edge of survival, they need to be assured that biochar really does make a difference to soil fertility. Third, they also need to be brought into the world's carbon trading systems. It is not enough to reward just commercial biochar plants in rich countries; we also need to remunerate smaller rural producers around the world.

The big problem, of course, is setting up a system that administers and monitors the scheme so that villagers in the tropics can get paid for sequestering carbon in the form of biochar, whether in tiny stoves or through the village's weekly charcoal-making session. No one denies that this challenge is substantial and perhaps even impossible. But the rewards are potentially so great, both in abating climate change and in alleviating poverty, that the global community needs to try to find a workable scheme. We will need an extensive monitoring organization, able to reward those farmers who maintain forest cover as well as check on improvements in soil carbon levels. Advances in satellite photography make it much easier than even ten years ago to check that the amount of land maintained as forest is growing as a result of the biochar program.

Speaking at a conference of biochar specialists in 2007, Australian Tim Flannery, one of the world's most respected earth scientists, fully endorsed their enthusiasm for using charcoal. "Your technology offers the possibility of taking carbon dioxide out of the atmosphere... and permanently sequestering that carbon in the soil," he said. "It does seem too good to be true, but I've looked at it from every angle, and I fail to see the fault in the system."

It would be very easy to dismiss biochar by saying that it requires too many new manufacturing plants or that the carbon storage in the soil is too difficult to monitor. But unlike most of the other carbon-reduction ideas in this book, biochar involves no great technological uncertainties or unknown costs. The problems are essentially managerial. And free markets are good at solving managerial problems. Biochar will need to be substantially bolstered over the next decade, but once established under the aegis of a reliable carbon price, it will support itself. The fact that biochar is also of disproportionate benefit to the poorer and rural parts of the developing world should make us even more enthusiastic.

SOILS AND FORESTS

Improving the planet's carbon sinks

WE TEND to portray the climate change problem as one-dimensional: the simple result of burning fossil fuels. This view is wrong. The issues are actually vastly more complicated, and, perhaps paradoxically, this complexity gives us greater reason for hope. In particular, the continuous circulation of carbon between the atmosphere, the oceans, soils, and plants provides us with the wherewithal to counteract the impact of fossil fuel emissions. Even if we can't quickly reverse the growth in fossil fuel use, we have some tools to offset these emissions by increasing carbon absorption elsewhere in the ecosystem. We can use carbon sinks to "swallow" some of the excess carbon dioxide arising from burning fuels.

In the last chapter, we examined one easy way of getting carbon dioxide out of the atmosphere and safely stored as carbon in the soil. This chapter also focuses on soil and, secondarily, on plants. The reason for this attention is that the world's soils appear to be the best available repository for excess carbon resulting from human action. As the chapter on biochar showed, much of the world's soil contains inadequate stores of carbon, and increases will benefit vegetation growth, which is also an important way of storing extra carbon. If we can make more food on existing arable lands, we don't need to cut down forests.

In contrast, increasing the oceans' carbon storage will be difficult. If we injected carbon dioxide into surface waters, the impact on the ocean's acidity would be severe, and anyway, the carbon dioxide would almost certainly return quite quickly to the atmosphere. There is already some worrying evidence that the capacity of the seas to hold the existing stock of carbon dioxide is reaching its limit. Injecting carbon dioxide into the very deep ocean where it would liquefy probably does not result in these problems, but it is technically difficult and would likely be very expensive.

So we need to do everything we can to improve the world's carbon sinks and lock as much carbon dioxide as possible into soils and plants. Adding charcoal to the soil works best for arable lands in tropical regions and areas of degraded soil. The biofixation techniques examined in this chapter will improve the carbon content of pastoral lands used for grazing animals and also add to the productivity of temperate croplands.

Finally, I look at the best and cheapest ways of reducing the rate of deforestation in the world. As forests are cut down and burnt, they give up their store of carbon to the air. Large acreages of woodland disappear every year, increasing greenhouse gas emissions as well as creating other severe ecological problems, such as desertification, soil erosion, flash floods in wet seasons, and drought in the drier portions of the year. Stopping the loss of trees may be the cheapest of all techniques for carbon capture and will also improve the long-term ability of the world to feed itself and support its rural populations.

Scientists have a rule of thumb that states that about half of the world's emissions stay in the atmosphere once put there by human activities. They disagree about how much of the rest is absorbed by the oceans and how much by forests, soils, and plants, as well as by the weathering of rocks. Probably more goes into the oceans than into land "sinks," but we cannot be completely sure. Some people think the amounts and proportions vary greatly from year to year.

But we can be reasonably certain that only about half of the carbon dioxide from fossil fuels and other sources ends up in the atmosphere. We can estimate both the weight of the earth's atmosphere and the weight of new carbon dioxide emitted every year. If all the carbon dioxide that the world produced ended up in the air, the atmospheric concentrations of carbon dioxide would rise by about 4 ppm every year, not the 2 ppm we estimated in the chapter on biochar. However, the evidence is increasingly strong that, little by little, the world's seas and lands are becoming less effective at capturing our greenhouse gases. A slightly larger fraction of what we emit today is ending up in the atmosphere rather than being safely sequestered. There's a tone of rising alarm about this subject in the major scientific journals.

The declining ability of the oceans and the land to capture our emissions will likely continue and possibly accelerate unless we take measures soon. For example, we know that the ability of the seas to take in carbon dioxide depends on the water temperature. Higher temperatures reduce the ocean's ability to dissolve carbon dioxide. Unsurprisingly, the temperature of the top layer of the world's water masses is generally getting hotter. So, all other things being equal, the amount of new carbon dioxide stored in the world's seas and lakes will probably fall.

This chapter examines the most effective and simplest ways in which we can retain and enhance the land-based stores of carbon—soils and forests—that would otherwise add to the inventory in the earth's atmosphere. Do these qualify as a "technology"? We are accustomed to applying this word to advances in electronics or perhaps in medicine, but it actually has a much wider meaning. One of the *Oxford English Dictionary*'s definitions is "a particular practical or industrial art," and it is under this rubric that I include the world's efforts of finding ways to productively store carbon in soils and forests.

GETTING THE MOST FROM THE LAND

A recurring theme of this book is that we will not be able to deal with global warming without addressing how we use the world's land. Land provides our food and, in most of the world, our cooking fuel, in the form of wood, vegetation, and dung. In a future low-carbon world, we will also be asking our soils to grow plant matter for cellulosic biofuels, district heat and power plants, and biochar. The urgent need to increase food production to feed an ever-growing population appears to be squarely in conflict with these new demands. We have seen the first, and very frightening, illustration of this conflict in the impact of first-generation biofuels on food prices.

Equally importantly, we must also improve the ability of the world's soils to retain carbon, and ideally to soak up more. These objectives appear to be irreconcilable. For example, more intense exploitation of the world's soils in an effort to produce more grain will probably decrease the amount of carbon that the land can store. If we devote a larger fraction of the earth's surface to growing trees or energy crops, the space available for crops will decline. How do we resolve these apparently incompatible objectives?

As I've already said, one obvious answer might be for the world to eat less intensively farmed meat. About 35 percent of the world's cereal grains get fed to animals. Even if nothing else changed, we could probably cope with a world population of 9 billion if none of us ate beef, lamb, or pork. Unfortunately, as the luckier half of the world population gets more prosperous, we tend to consume more meat, increasing the amount of grain eaten by animals and therefore reducing the supply available to those who rely on cereal grains as their primary source of calories. For most people, consuming more animal protein is an important benefit arising from increased prosperity. For that reason, cutting the link between increasing wealth and industrial meat production will prove very difficult, so I don't propose this as a solution to the conflict between our food needs and the requirements for low-carbon energy sources.

In fact, later in this chapter, I suggest that better grazing practices may allow greater amounts of animal husbandry in many areas of the world. Although herbivores will always be significant producers of methane (created as a by-product of the digestion of cellulose), the little research on this topic suggests that animals kept outdoors and eating only grasses produce far less of this greenhouse gas than factory-farmed cattle. They also do not eat cereal grains and remove the grains from the food supply. In reality, then, it is not meat that is the problem—it is industrially farmed meat.

Instead of simply concentrating on meat production, we should focus on maintaining and improving the level of carbon in the world's soils. As the previous chapter showed, the world's soils contain about twice as much carbon as is in the air. So small changes in soil composition can result in significant flows of carbon into or out of the atmosphere. Agricultural practices around the world must be geared in part to maintaining and improving the ability of soils, whether cropland, pasture, or forest, to hold carbon.

ZERO-TILL

Intensive cereal farming tends to reduce the amount of carbon in the soil through a variety of mechanisms. Plowing the soil exposes organic matter such as humus to the oxygen in the air and makes it likely to rot. Exposing the bare soil to rainfall after harvest increases erosion of the normally carbon-rich top layer. Removing straw and other waste matter similarly cuts the amount of carbon left in the soil. These outcomes are bad for the farmer, because the complex carbon-based molecules in soil humus are invaluable for maintaining the structure of the soil and for making nutrients available to plants. They are also bad for the planet because the carbon in the soil degrades into molecules such as carbon dioxide and methane, which then add to the stock of greenhouse gases in the atmosphere. All things being equal, soils with high levels of complex carbon molecules will be good for the climate and for agricultural productivity.

In tropical soils used for growing crops, biochar will help improve soil carbon levels. On temperate cereal-producing soils, however, the most effective way of increasing soil carbon levels may be to shift to a form of agriculture known as "zero-till." Farms using this approach never use a plow to turn over the soil. The proponents of the zero-till technique believe that plowing is counterproductive because it reduces the soil carbon content and increases the loss of valuable moisture. Perhaps surprisingly to those of us who love the smell and appearance of a freshly turned cereal field, plowing has little direct benefit to the crop. It reduces weed growth and helps expose some harmful insects to birds and other predators. But these benefits are offset by the potentially detrimental effects on the soil.

Zero-till farmers disturb the soil as little as possible. They plant the seeds in a row, accompanied by fertilizer. Once the main crop is harvested, a second crop is planted to provide cover for the soil to prevent erosion and any loss of fertility. Often called "green manure," this cover crop is cut down and left on the surface of the soil to slowly rot and fertilize it. These dead stalks and leaves help protect the soil from erosion caused by heavy rains and provide a rich source of organic material that earthworms can use to improve the quality of the lower soil. The following year, seeds are planted through this material, known to farmers as "trash." Regularly varying the crop prevents the buildup of destructive pests and diseases.

The evidence that zero-till farming increases soil carbon levels, partly by minimizing erosion by water, is now compelling. It may not work on every type of soil, and results tend to vary from year to year, but switching from conventional arable farming techniques to zero-till seems to reverse the trend for heavily cropped soils to lose carbon. In addition, crop yields seem to improve, although for many farmers there is an initial period during which productivity suffers. In other words, although the debate is not yet over, zero-till could increase the tonnage of grain the world harvests every year *and* turn intensively farmed soils from being net emitters of carbon to being an increasingly valuable store.

Of course, the zero-till approach has its drawbacks, at least in some people's eyes. It uses large amounts of herbicides to control weeds that would otherwise have been averted by deep plowing. In fact, zero-till may work best with crops that have been genetically modified to withstand the most common weed-killers so that herbicide can be applied while the crop is growing. Either way, the zero-till approach is largely incompatible with organic farming, which bans the use of genetically modified seeds and herbicides. Organic farming, for all its potential benefits, uses regular plowing to ensure that weeds do not become dominant, thus tending to reduce the ability of the soil to store carbon.

From a global warming point of view, organic farming has another disadvantage. Many organic farms produce cereal yields of little more than half the level of conventional farms. The difference largely arises because organic farming builds up fertility by planting nitrogen-fixing plants such as clovers for several years after each grain crop rather than using artificial fertilizers. Put another way, if the entire world's cropland were farmed organically, we might need twice as much land to feed the global population. Organic farming systems usually also require a large number of animals to provide the manure that fertilizes the cereal fields. In many countries, these animals need substantial amounts of winter feed. These aspects of organic farming all put pressure on the world's limited stock of high-quality farmland, increasing the incentives to cut down forests.

However, organic farming also has substantial advantages. For example, it helps reduce nitrous oxide emissions because it avoids using artificial fertilizers. And the long crop rotations probably help restore the soil carbon losses caused by frequent plowing. The point I want to make is not that organic farming is bad but rather that zero-till techniques may be better at simultaneously providing high yields of foodstuffs and maintaining soil carbon levels.

As a result of the large advantages zero-till farming offers on most soils, a large percentage of total arable land in some parts of the world has now been converted to zero-till cultivation. Over 20 percent of American farmlands are now avoiding the plow, and the figure is even higher in Brazil and Canada. Zero-till has made slower progress outside the Americas and Australia, perhaps because in Europe and other high-latitude zones low levels of soil carbon are not yet a threat to production levels. Soil carbon losses in conventionally tilled soils are rapid only above temperatures of about 25°C (77°F), a threshold reached only for a few days each year in high latitudes. Nevertheless, there is increasing evidence that even in cooler countries, zero-till will significantly help maintain or increase the amount of carbon stored in arable soils.

In Brazil, earlier phases of intensive cultivation may have cost 30 or 50 percent of the original soil carbon in the dry "cerrado" region south of Amazonia. Although research produces very varied results, a typical acre of soil in this area, if cultivated with zero-till techniques, absorbs 0.2 to 0.3 tons of new carbon each year, equivalent to roughly a ton of carbon dioxide, suggesting that if the grain-producing area of the entire world (about 1.7 billion acres) were switched to zero-till, the sequestration might be worth almost 2 billion tons of carbon dioxide a year, between 5 and 10 percent of today's total emissions. Perhaps the impact in higher latitudes would be less than it is in Brazil— we really don't have good information on this yet—but a robust global climate change policy would probably include providing incentives to use zero-till techniques. In addition, a switch to zero-till techniques would reduce the amount of fossil fuel used on farms because tractors would not be used as much. The extra use of herbicide would only partly wipe away this gain.

Much as we saw with biochar, a further benefit of zero-till may be the reduction in nitrous oxide emissions from soil. On a zero-till farm, smaller amounts of fertilizer may be used, sometimes placed next to the seed in the sowing process. Researchers are not yet completely confident, but it seems that the limited and highly targeted application of nitrogen fertilizers in a zero-till system may potentially provide another significant advantage over conventional farming methods, which spread the fertilizer over the whole surface of the field during the growing season.

Along with the other techniques described in this chapter, a huge increase in the use of zero-till farming may be an excellent way of keeping carbon in intensively farmed soils. In the long run, however, there may be an even more effective way of improving the world's carbon sinks: reviving degraded grasslands.

REVIVING GRASSLANDS

Australian Tony Lovell isn't a farmer, but his attempts to demonstrate the importance of increasing the soils' carbon levels have begun to attract attention. He persuasively argues that the easiest and best way to reduce the excess carbon dioxide in the earth's atmosphere is to improve the soil health of dry pastoral lands. Overgrazing and erosion over the decades have taken much of the carbon out of most of the world's enormous areas of ranchland.

Lovell is an accountant by training, and within a few minutes of starting a conversation with him, you get drawn into doing some ecological arithmetic. But first he shows you some photographs, starting with a ranch in Mexico: empty, bare soils, with very little vegetation. Next is a photo of the adjacent property, a few miles away. Same soils, same rainfall, but abundant growth and a healthy ecosystem supporting cattle and a range of wildlife. The difference is astounding. What causes the huge variation? According to Lovell, it's simply a matter of how animals graze the land.

Before people came along, Lovell explains, huge herds of herbivores such as bison or wildebeest moved across the plains of the Americas and Africa. The animals stayed together in a group because they feared predators such as lions. They kept moving because otherwise there would not be enough grasses for the large group. Once a stretch of grassland had been eaten, the herd would not return to the area for several weeks, by which time the crop would have grown back.

Humans changed all of this. The herbivores' predators were gradually killed, allowing the animals to roam singly or in small groups. Liberated from dangerous carnivores, these animals could repeatedly feed on the same areas, keeping plant growth to a minimum and allowing no long recovery periods for the grasses.

Lovell explains the importance of this shift. Unlike a tree, he says, most of the weight of a grass lies under the ground. Whereas trees keep their stored carbon in their trunks and branches, the carbon captured by grasses ends up mainly in their large root systems. The roots of some grasses can go several feet into the ground. But when the grass is eaten and its height is reduced, the plant sloughs off some of its root system. Within minutes, Lovell says, a grass that has had much of its leaf system removed by a grazing animal will start to reduce the length of its roots. This previously stored carbon gradually returns to the atmosphere as the dead matter rots away. If the grass is then not eaten for a period, the roots will be gradually reestablished. But grasses that are repeatedly cropped will never again build up long, carbon-rich roots, cutting the amount of carbon in the soil. Partly as a result of this process, the ground will hold less water, and almost all vegetation will eventually die. Enhanced rates of erosion arising from the lack of roots to bind the soil together and the absence of vegetation to protect the surface complete the vicious circle. Occasional flash floods remove most of the carbon that remains.

Pastoral soils that have lost most of their carbon can be seen all over the world but particularly where the rainfall is highly seasonal. Lovell calls these areas "brittle," an expression first coined by his mentor, the great Zimbabwean pastoralist Allan Savory, who now runs a center in the U.S. seeking to improve the management of the world's vast rangelands. A brittle area has long periods, of perhaps six months or more, in which virtually no rain falls. In non-brittle areas, such as the croplands of northern Europe, misuse of the land does not result in a desertification: the regular rain allows the land to partly recover. But in brittle regions, the damage caused by overgrazing will usually tip the land into what is generally thought to be an irreversible decline.

Poor grazing practices—sometimes imposed a hundred or so years ago by new immigrant farmers from water-rich regions—helped reduce the carbon in the soil. These practices also damage food production, by massively reducing the ability of the land to feed animals. The impact on poor rural communities can be devastating. For example, in recent years, the threat from ever-growing deserts in China has forced the government to ban animal grazing over vast tracts of land in the west of the country. By completely stopping any use of the land by pastoral animals, the Chinese are trying to help restore the health of the soil, increasing its ability to hold water and grow grasses with deeper roots. This may well have been a necessary step, but it has adversely changed the lives of many thousands of pastoralists. Unless stopped by draconian measures such as these, overgrazing will eventually help create deserts in many of the driest parts of the world.

According to Lovell, Savory, and others, reversing the process that is degrading huge expanses of land around the world is difficult but manageable. The task is to recreate the way herbivores used to crop the land—in large groups that move swiftly across the landscape, not returning until the grass has recovered. So instead of allowing animals to range in small groups across the entire landscape, the herbivores need to be herded carefully. A sheep farmer might move the livestock every four or five days from enclosed pens, not coming back to the same place for ten weeks. Or cattle could be moved by herders opening and closing water points and actively herding the animals from one place to another.

Lovell points out that grazing itself is not the source of the problems. Through their dung and urine, the animals return almost all the carbon and nutrients taken from the soil and, paradoxically, by trampling down plants, the beasts help to maintain a layer of decomposing organic matter that protects the soil. As Lovell and others are at pains to point out, proper grazing management can actually substantially *increase* the number of animals that a ranch can support. An "overgrazed" landscape is not one that has had too many animals on it but one on which grazing is not properly controlled. By contrast, good management means water is retained and grasses grow quickly. Patches of animal manure provide places for new seeds to safely germinate and help build the soil.

Some scientists have known for generations that controlling the movement of pastoral animals is crucial to maintaining soil health in brittle lands. One of the classic books about the human influence on climate put this concept very clearly in 1977. In *Climates of Hunger*, Reid Bryson and Thomas Murray wrote:

> A fence built round a large field in a desert brought forth abundant wild grasses in two years, without planting or irrigation, mostly by keeping out the men and the goats. As similar experiments in the western U.S. show, grasses do grow if grazing pressure is lifted.

In fact, vegetation on properly grazed land, fertilized by manure, will often grow faster than vegetation on completely abandoned land. One of Lovell's most extraordinary photographs shows an enormous 1,000-acre pile of copper-mine tailings in Arizona. After sixty years, the massive hill had virtually no vegetation cover. Eroded gullies ran down the side. Then cattle were introduced on part of the pile. Fed with hay, the animals excreted carbon-rich manure that gradually provided an environment in which plants could establish themselves. The animals' feet pushed remnants of their feed, covered in fertile dung and urine, into some of the holes in the surface, creating good environments for grass growth. Eventually the cattle created soil a foot thick. The soil substance is, of course, almost entirely organic matter that has grown by sequestering atmospheric carbon dioxide through the photosynthesis process. The roots of the grasses have also bound the surface of the mound together, meaning that erosion has stopped and the carbon will be permanently stored, provided the careful grazing regime continues.

Another photograph shows an adjacent area that was kept free from animals and instead treated with grass seeds and regularly watered. This attempt to stabilize the mound and build a resilient soil completely failed. In the picture, the surface is still completely bare and scarred by deep gullies.

Photographs like these make an utterly compelling case. So why hasn't the world rapidly adopted the advice of Allan Savory and his followers? The answer to this question is partly that any form of farming innovation is highly risky. If things go wrong, the farmer could be ruined. In addition, many farmers already live on the edge of economic failure (partly as a result of falling income caused by bad grazing practices). In these circumstances, the tendency to resist new ideas is understandable. But once a few farms move to carbon-enhancing grazing management, many of their neighbors eventually follow. Of course, there may be other problems as well. Just as I finished a draft of this chapter, I met a distinguished Namibian architect whose father had followed Savory's principles in managing the family farm in that dry and hostile climate in southern Africa. She agreed that soil management was vital and had helped her father grow cereal crops in an area where water shortages and erosion made arable agriculture very unusual. But one year, a severe fire had overrun part of the land. The farm's sheep had all been fenced into a small area, in line with Savory's recommendations, and were unable to escape the blaze. Grazing practices that allowed the sheep to roam freely would have prevented this severe blow to the farm's fortunes.

Despite problems such as these, better grazing management will generally be good for farm incomes and also for biodiversity and soil fertility. What about the impact on global warming? Tony Lovell's figures suggest a remarkable answer. Improving grazing lands may be the single most important step we can take to combat climate change and to feed the world. Lovell says that the world has about 12 billion acres of brittle pastureland, perhaps seven times the area of land devoted to crops, almost all of which could be managed for improved food production and enhanced carbon storage.

Let's look at what happened when one thoughtful farmer switched to better grazing management. Lovell quotes figures from a pioneer in the dry Gulgong area of the New South Wales tablelands whose 2,200 acres grow some wheat but also provide the grazing land for over four thousand sheep. This innovator increased the carbon in his farm's soil by over 50 percent over a ten-year period by carefully improving his grazing practices and by planting his wheat in rows interspersed with pastureland. The average level of soil carbon rose from under 2 percent to over 3 percent of the mass of the soil.

This sounds like a small difference. It is not. Making this apparently insignificant improvement to the top foot of the soil sequesters a massive 40 tons of carbon dioxide per acre. It also substantially improves the productivity of the land and makes it more resistant to the impact of drought, an extremely important consideration in Australia, where climate change is likely to cause long-term water shortages over much of the continent. In fact, most of the currently dry areas of the world are likely to be at substantially enhanced risk of catastrophic drought in our hotter world.

The experience of this pioneer and many others is that after a few years of transition, the profitability of farming is improved, just as with zero-till cultivation. In other words, this practice is a very low-cost form of carbon sequestration. Like biochar and growing algae, it may be a much cheaper form of carbon capture than the techniques for capturing emissions from power stations described in Chapter 8. It also has the profound advantage of being ready to roll out, unlike carbon capture technologies that are still being developed.

There are about 490 million acres of brittle grazing land in Australia alone. So if we could find a way of adding 1 percent to the carbon content of all these soils, we would be storing almost 20 billion tons of carbon dioxide, about 80 percent of the world's current annual emissions. Similar dry ranchlands can be found across the tropical and subtropical world, including much of sub-Saharan Africa and large areas of South America, as well as the southwestern U.S. and parts of Mexico. It would be foolish to suggest that a single management approach will work in all these regions. What succeeds in countries with mobile herders, such as in some parts of Africa, Asia, and even southern Europe, will not succeed in more settled pastoral communities or in areas that are almost desert. The amount of extra carbon that can be stored will also vary. But the potential is huge: if 1 percent extra can be achieved across just half the world's brittle lands, that would soak up the world's current emissions for over ten years.

There is a very strong argument for including soil sequestration in the European carbon market, and any future U.S. or global systems. The carbon dioxide moved from the atmosphere to the soil as a result of better land management has a value to the global community equivalent to a reduction in power station emissions. A farmer storing an extra 40 tons of carbon dioxide in an acre of land should be able to receive a reward of $800 for his or her valuable effort. For farmers struggling to survive on increasingly degraded land and facing the prospect of more erratic rainfall as a result of climate change, this amount of money is a vital addition to farm income. A 2,500-acre ranch, perhaps on the Argentinian pampas or in the southwest U.S., could earn over $3 million by increasing soil carbon levels by 1 percent over a ten-year period.

Of course, we can expect abuses of a system like this, and it would be troublesome and complex to administer, but not necessarily more so than the Clean Development Mechanism (CDM), one of the Kyoto Protocol's carbon trading tools. Many projects that have qualified for payments under the CDM have turned out to be little more than hoaxes, but soil sequestration schemes are less likely to be misused. After all, storing carbon this way is cheap, technologically simple, and easy to validate. Success might mean eventually returning the amount of organic matter in the soil to the levels prior to human intervention. The benefits of doing so would be much wider than just climate change, with positive implications for food supply, water availability, and the incomes of struggling farmers.

Scientists have advanced many different high-tech solutions for dealing with the existing stock of CO_2 in the atmosphere. As we'll see in the epilogue of this book, many of these ideas are expensive or potentially dangerous or both, and the results are uncertain. By contrast, Tony Lovell's vision is simple and virtually risk free: to return the earth's pasture lands to their condition prior to the not entirely benevolent takeover by humans. The side-effects of improved grazing management, such as improved biodiversity and productivity, are likely to be positive.

Along with zero-till farming, better grazing management shows that there needn't be any conflict between our climate change and our agricultural objectives.

FORESTS

Many of us realize that the world is rapidly losing its forest cover. In particular, we all understand the threats to the Amazonian rainforest, under continuous pressure from illegal logging and conversion to soy plantations and cattle ranches. Similarly, Indonesia is changing virgin forest to huge areas of palm oil plantations. Deforestation in these countries and elsewhere is increasing the pace of global warming by adding carbon to the air.

Although forest loss is a hugely significant issue that has captured media attention, the reality is possibly slightly less bleak than most people imagine. Although the world is losing its forest cover, the rate of loss is probably now declining. The UN Environment Programme (UNEP) says that global deforestation amounts to about 32 million acres a year, out of a total of about 10 billion acres of woodland, or somewhat less than 0.5 percent a year. This reduction is partly compensated by the creation of new forests, although no one should pretend that establishing new plantations compensates for the many detrimental effects of the destruction of virgin woodland.

That point aside, we can estimate that the net loss of forested area is probably between 17 and 20 million acres a year, or 0.2 percent of the world's wooded area. Forests are actually increasing in size in Europe and in some other parts of the world. Largely to try to prevent further desertification of large parts of its interior regions, China has planted over 50 billion trees in the last two decades or so, creating 130 million acres of new forest. This step has helped slow the march of the deserts that cover over a quarter of the land area of this huge country. The success of the Chinese investment in trees proves that deforestation is not an inevitable process that the world can do nothing about. Well-organized countries can improve the level of tree cover. Nevertheless, deforestation is still playing an important part in increasing the levels of greenhouse gases in the atmosphere and requires us to respond urgently.

Wood is approximately 50 percent carbon. When forest is lost, this carbon is largely transferred to the atmosphere, particularly if the wood is burnt. UNEP says that about 4 billion tons of carbon dioxide, containing over a billion tons of carbon, are lost from forest biomass every year, equivalent to perhaps 15 percent of the total additions of carbon dioxide to the atmosphere. Scientists are right to repeatedly draw attention to this huge figure. But the optimists among us need to remind ourselves that new, carefully managed wood plantations in the humid tropics could completely replace this annual loss by the natural yearly growth of an extra 200 million acres and would involve adding only 2 percent or so to the world's forested area or less than twice the amount the Chinese have already created. Forest plantations may not be to everybody's liking, not least because they tend to use a small number of species and introduce limited biodiversity of animal and plant life. But a well-managed planting program in the tropics will foster trees that grow ten or twenty times as fast as unmanaged woodland in the temperate zone.

By intelligently planting the right tree species in woodland plantations where growth is likely to be rapid, we can roll back some of the appalling effects of forest loss in the Amazon and elsewhere. This point must come with a caveat. The gradual loss of the Amazonian rainforest will eventually not just affect greenhouse gas levels but will also change the world's weather patterns, probably making large areas in the western hemisphere much drier than they are now. By stressing that an active program of reforestation is a vital ingredient in combating climate change, I don't want to give an impression that I think we can simply let the Amazonian and Indonesian logging continue.

Deforestation in these areas looks like an intractable problem. Some forest clearing is occurring as a result of large commercial farmers wanting to add to their estates. But it also happens partly as a result of the decisions of millions of people trying to create new land on which to grow food. In many countries, land tenure law does not give clear ownership rights over these trees, making protecting the forest very difficult. States such as Brazil may be able to restrict the loss of land to soy crops and beef ranching, but telling the landless poor that they should not try to feed themselves by growing crops on cleared land is a policy that is unlikely to succeed anywhere in the world.

However, the longer-term effect of widespread woodland loss is to increase soil erosion and cut fertility. It will also probably change local microclimates, eventually reducing total agricultural production. The chapter on biochar explained how tropical slash and burn techniques could be amended to keep some proportion of newly cleared wood as carbon in the soil, helping to improve fertility and reduce erosion. This practice will help reduce soil carbon losses.

But we can do other things as well. Probably the single most important change we could make would be to reduce the amount of wood cut down for cooking fuel. Energy-efficient biochar stoves would help, but many other types of improved stove design have been tested throughout the world. Some are made from ceramic or brick components and others from metal. By capturing more of the heat from burning wood, they reduce the amount of firewood needed and therefore cut the amount of local forest loss. Particularly in Africa, where UNEP says that half the net loss of forest is caused by the need for cooking fuel, efficient stoves could significantly reduce the rate of deforestation. As we've seen, they also burn more cleanly, preventing dangerous local air pollution.

Of course, what looks simple and economical on paper does not always work quite as well as expected in the field. Studies looking at the roll-out of new stoves have shown that although the savings in wood fuel use might be as much as 50 percent, this target is often not achieved. People don't always operate them in the most effective way—or even abandon them because they take longer to cook food. Hungry people don't want to wait any longer than necessary to eat.

Nevertheless, a worldwide effort to improve the efficiency of cooking remains crucial to addressing the forest loss problem. At the moment, the drive to reduce the amount of wood used in stoves is being led by a mixed bag of small charities and "carbon offset" specialists (organizations that establish carbon-reduction projects to counterbalance emissions caused by their customers' air travel and other activities). One example is Climate Care, a subsidiary of the U.S. bank J.P. Morgan, which spends much of its income developing and marketing new types of stoves in countries as different as Honduras and Bangladesh. This work is a very useful start, but the world's richer countries should contemplate a much larger and more widespread effort to reduce the number of trees cut down for cooking. About 2 billion people (nearly a third of the world's population) cook using wood or dung. We cannot really be sure of these figures, but some estimates suggest that each person needs about 1,700 pounds of cooking fuel each year. Multiply the two numbers together, and we get a global figure of about 1.6 billion tons of wood and dung used each year, of which about 800 million tons (50 percent) is carbon. UNEP says that forest loss produces about 1.1 billion tons of carbon a year, meaning that cooking accounts for about three-quarters of the net loss of forest volume. Put another way, if we could cut the amount of wood fuel used in cooking stoves in half, this step alone would reduce net human-made additions of carbon dioxide to the atmosphere by over 5 percent.

Fuel-efficient stoves are one of many ways of reducing wood fuel use in tropical countries. Other agencies are working on solar cookers, including various devices that focus the sun's rays onto a container of water. These solar cookers use exactly the same principle used in the concentrated solar power plants described in Chapter 2: using mirrors to reflect sunlight onto a small area generating intense heat.

Another promising technology is the biogas collector. When human, animal, and plant wastes rot down, they give off methane, particularly if kept in enclosed containers or left in a garbage dump or landfill site. In most parts of the world, this methane vents to the open air, but if it is collected, it can be burnt in simple stoves, replacing the need for woody fuels. (What we call natural gas is almost pure methane.)

Alongside its policy of large-scale reforestation, the Chinese government has heavily promoted biogas stoves to reduce the need for families to burn wood. In rural areas, concrete or brick-lined pits next to houses collect latrine waste and other materials that will rot, such as straw from fields. The pit is covered, and sometimes a greenhouse is put on top. The greenhouse is warmed by the heat from decomposition, and, in turn, the greenhouse helps insulate the pit so that temperatures can be high enough for the rotting process to continue. The resulting methane is taken by pipe to a cooking stove and sometimes to gas lamps. When the decomposition has finished, the waste matter, now sterile, can be used as fertilizer to fields or even fed to pigs. In the past, human wastes might have been put directly on to fields, from where pathogens could flow into local watercourses, so the new approach also offers substantial health benefits.

A biogas cooking stove may not completely remove the need for other sources of energy for cooking. In times of low ambient temperature, for example, the rotting process will slow down and the amount of methane produced will taper off. At such times, the house may still need wood fuels for cooking. However, in most of rural China, average temperatures are high enough to produce enough methane for at least six or eight months a year.

Official figures suggest that over 18 million Chinese households now have biogas stoves. Estimates of the amount of firewood saved can be little more than guesses, but the figure could be as high as 30 or 40 million tons every year, equivalent to up to 2 percent of the total worldwide loss of forest carbon. The Chinese government says that the number of household biogas digesters can be increased eightfold, implying further huge potential reductions in the use of wood fuels as well as improved soil fertility and lower rates of waterborne disease. As with the tree-planting program, the campaign to install digesters to decompose waste in China has been driven more by the need to prevent deforestation than to reduce greenhouse gas emissions. But the two go hand in hand.

As well as encouraging the installation of domestic biogas systems, the government has helped set up waste-digestion systems in animal-rearing units. These larger-scale anaerobic digesters then pipe methane—again for cooking and lighting—to local villages. Previously burnt or simply thrown away, agricultural straws and stalks are also increasingly digested in these methane-producing tanks. Copied across the world, these cheap and simple technologies can systematically reduce the amount of wood needed for cooking.

My aim in the last few pages is not to suggest that deforestation is a simple issue to address. It is not. The pressure in some parts of the world to cut down trees for fuel or to clear agricultural land is intense. Eventually, as the Bali climate change conference decided in 2007, the world will have to pay local people to maintain forest lands, particularly in areas where population pressure is at its most intense. But until we have put in place a system of appropriate financial incentives, the right thing to do may be for the rich world to focus on the three simple technologies just discussed: fuel-efficient stoves, solar cookers, and, perhaps most importantly, biogas tanks.

Each of these measures reduces the need to cut down trees, and biogas digesters, by producing fertilizer, may also increase production from arable lands. As with biochar, zero-till, and improved grazing management, it's a way of saving carbon while simultaneously boosting food production. These techniques are the simplest and most basic of the portfolio of solutions to climate change, but they may also end up being the most effective.

As a footnote, it's interesting to note that while China has over 3,500 biogas plants treating animal wastes on commercial farms

270

(in addition to the 18 million domestic units), the U.S. has only a few hundred. Wisconsin, for example, has only about twenty-five digesters, despite being a center of livestock production. Anaerobic digestion isn't particularly complex or expensive, and we need to use this technology wherever cows and pigs are reared.

PUTTING IT ALL TOGETHER

Are the ten technologies enough to save the planet?

POLICY-MAKERS AROUND the world believe that in order to keep the global temperature increase below 2 degrees Celsius (3.6 degrees Fahrenheit), we must ensure that atmospheric concentrations of greenhouse gases eventually stabilize at no more than 450 ppm. This means that the world's industrialized nations will need to reduce their emissions to 20 percent of today's levels by 2050. Some people say that the reduction will have to be even greater.

The rate at which this cut is achieved can vary. We might choose to wait until 2020 before reducing our emissions. But leaving it this late would mean extremely fast annual decreases would then be necessary. The global consensus is that the appropriate way to get emissions down to 20 percent of the current level is to halt the rise in annual emissions soon and then pursue a less rapid reduction plan. But this is not going to be easy. Even if we ensure that the emissions of the industrial world hit their maximum in 2010, we will need to cut emissions by 4 percent a year thereafter if we are to reach the 2050 target.

If we make substantial progress in each of the ten

opportunities described in this book, will we be able to reduce fossil fuel use in advanced economies fast enough? And can we counterbalance the emissions of methane and other greenhouse gases by improving carbon storage in soils and in plant matter? The best way to address these questions is to examine how energy is used in advanced economies and then estimate how much can be switched from fossil fuels to green sources. If progress is good, where might we able to get to by 2025?

Let's look first at how much fossil fuel energy the modern world uses. We can calculate the total energy use of a country by adding up the volume of fuels used, multiplying by their energy content, and then working out a figure for each inhabitant. The numbers are large. In a typical industrialized country, around 50,000 kilowatt-hours of energy are used to support the lifestyle of each person a year. The figure is approximately twice as high in North America, and is lower in Japan. The fast-growing countries of Asia and elsewhere are lower still, at perhaps a quarter of the European level.

To express this figure another way, 50,000 kilowatt-hours per year is equivalent to a continuous stream of 5,000 or 6,000 watts—comparable to two electric kettles boiling day and night.

Some of this energy use is obvious to us—for example, operating our electric appliances—and some is happening invisibly in the factories and offices that supply us with goods and services. In most industrialized countries, almost half of all energy use is a direct consequence of how individuals run their lives. The gas used to heat homes, the fuel for cars, jet fuel for airplanes, and the electricity for home appliances account for about 45 percent of total energy use. Commerce uses another 20 percent, and industry accounts for the final 35 percent.

274

ELECTRICITY

The proposals in this book will add to electricity demand. I estimate that the total production of power will need to rise by about 50 percent to meet the extra requirements of electric cars and vans, the increased use of electricity for heating and cooling, and the replacement of oil and gas in some industrial processes. (Many industries will continue to need fossil fuels for the most energy-intensive processes, such as smelting ores to make metals.) As Figure 1 shows, this increase will take typical electricity use to about 2,750 continuous watts per person, up from about 1,900 today.

FIGURE 1: *Possible increases in electricity demand*

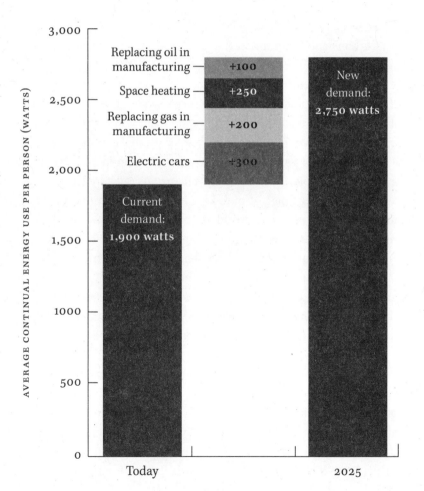

How will this electricity be generated? By 2025, I believe it will be possible to have a mixed portfolio of renewable sources of power that provide most electricity without carbon emissions. Carbon capture technologies will stop the carbon dioxide from the remaining gas and coal plants reaching the atmosphere.

While these figures can only be guesses, how might this portfolio look?

Wind power	25 percent
Solar power (mostly CSP in hot deserts)	25 percent
Marine (tidal and wave)	15 percent
Fuel cells and biomass CHP	10 percent
Carbon capture (or nuclear)	25 percent

Are these numbers feasible? Let's look first at wind power. If we want 25 percent of the (increased) demand for electricity in industrialized countries to come from wind, this means about 700 watts of continuous power per person, or about 6,000 kilowatt-hours a year. This will require a big wind turbine for every 1,800 people. If the turbines were in good offshore locations and as large as the biggest of today's models, we'd still need one for every 3,000 people. If world electricity usage rises to North American levels, the numbers of turbines would be twice as large.

Clearly this is a substantial challenge, but it is far from impossible. If all the turbines were put offshore, they would fill an area of about 60 miles by 100 miles. This is only a small percentage of the shallow waters off North America. It's worth mentioning that in some parts of Europe almost 25 percent of electricity demand is already met by wind.

Solar power, primarily produced in large-scale concentrated solar power plants in deserts, could also comfortably produce about 25 percent of North America's electricity needs and a similar percentage of Europe's. As Chapter 2 demonstrates, the primary obstacle is the need for a sustained program to construct long-distance DC transmission lines. By contrast, the growth of electricity supply from tides and waves depends on continued entrepreneurial activity among the plethora of small firms constructing innovative devices. But for regions with long west-facing coasts, such as the Pacific coasts of Canada and the northwestern U.S., a target of 15 percent of electricity supply is attainable. Meeting the fuel cell target also depends on technical progress, particularly in reducing the cost of cells designed to produce electricity for large commercial buildings. The use of biomass for combined heat and power needs few technological advances but is dependent on the ready availability of woody materials to use as fuel.

Generating this much power from renewable sources will require us to rethink the way we run our electricity grids. But, as discussed in Chapter 1, this obstacle is not insurmountable. We just need to develop more energy storage capacity, internationalize electricity grids, and find ways to reduce peak demand.

The remaining electricity demand is about 700 continuous watts per person, or 25 percent of the total need for power in the median industrialized country. If, as is likely, we continue to use fossil fuels to generate this electricity, we will need to capture the carbon dioxide that results. ccs technology, as discussed in Chapter 8, is a long way from commercial availability. Indeed, it may be 2020 before we understand how to capture CO_2 with reasonable efficiency. But then it should be possible to add carbon capture equipment to most existing coal-fired power stations. The world may also decide to invest in large numbers of new nuclear power stations. In combination, nuclear and ccs-equipped conventional power stations will be able to produce the quarter of our electricity needs that does not come from renewable sources.

277

OTHER GAS, OIL, AND COAL USE

Completely decarbonizing electricity production by 2025 is a tough challenge, but we can clearly see the routes that we need to take. Less work has been done on reducing gas demand. Gas that is not burnt to generate electricity is primarily used for heating buildings and to provide heat for industrial processes such as food manufacture.

Improvements in building insulation are the most effective way of reducing gas demand. The chapter on super-insulated homes shows that heat needs can easily be cut in half by relatively cheap improvements in insulation. Progress in cutting gas use in the home is fastest in places like Germany, where very substantial insulation improvements are benefiting several hundred thousand households a year. But even this figure means that only about half of a percent of all existing households in that country are undergoing major eco-renovation each year. The world needs to step up the rate at which insulation improvements are being made, but there are no technical obstacles to this, either in cold or in hot countries.

As Figure 2 shows, further emissions reductions will come from increasing the amount of heating provided not by gas but by low-carbon electricity. We can also employ renewable fuels to provide heat, either through fuel cells powered by next-generation ethanol or by wood in district heating plants. Fuel cells for homes and for larger buildings will almost certainly be financially viable within a decade, particularly if today's gas prices increase and the district-heating model that has already been proven in Denmark and elsewhere is widely adopted around the world.

Most oil is refined into diesel and gasoline and used as a fuel for transport. In Chapter 6, we showed how liquid fuels can be replaced by electric batteries in cars and vans. Other vehicles can be powered by cellulosic ethanol, as described in Chapter 7. We will not find it as easy to replace kerosene aviation fuel or diesel for heavy vehicles. Although diesel substitutes can be made from any oil-bearing seed, biodiesel is problematic because its production generally involves switching either virgin forest or food-producing land to land for energy crops. We can hope that a large percentage of fossil fuel diesel can be eventually replaced

FIGURE 2: *Measures to reduce gas demand*

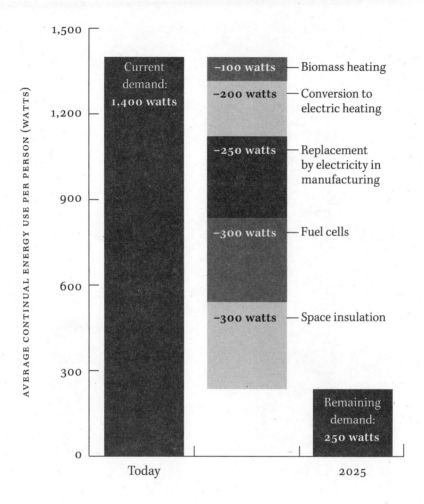

by fuel made from dried algae, as described in Chapter 8, but it is not yet mature technology.

Oil currently provides about 1,650 watts of continuous energy in a typical industrialized economies. About half of this energy is used in the fuel that drives cars and vans. Figure 3 shows how we can hope to substantially reduce the amount of fossil fuel but still use approximately the same amount of energy.

FIGURE 3: *Measures to reduce oil demand*

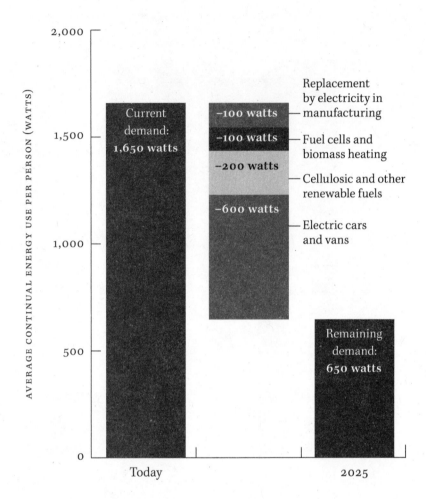

In this possible 2025 scenario, oil use falls by 600 watts as the electric vehicle takes over. (Electricity use only rises by about 300 watts because electric motors are more efficient at converting energy into motion than the internal combustion engine. The same amount of driving requires less energy.)

Even after the possible reductions detailed in this book, we are left with 650 watts of power demand fulfilled by oil. This figure is composed largely of aviation fuel (about 300 watts), shipping (100 watts), heavy road transport (200 watts), and a small amount of power for industry that can't easily be replaced with power from renewable sources.

Lastly, we need to consider coal. Excluding the use in electricity generation, coal provides about 150 continuous watts in advanced economies, only a small proportion of the total. Coal is generally used for heavy industrial purposes, such as in blast furnaces to make iron, and these uses will be largely unaffected by the technologies discussed in this book. The figure will also vary substantially country by country depending on the type of local industry. In some countries, coal is still used to provide a substantial portion of home heating. In these places, we can hope that it will be replaced by renewable electricity.

Adding together the remaining needs for gas, oil, and coal, the total need for energy from carbon sources could therefore fall to as low as 1,050 watts by 2025, down from over 5,000 watts today. With a following wind, the ten technologies in this book could therefore reduce fossil fuel energy needs in a typical advanced economy by up to 80 percent. This figure conveniently matches most scientific assessments of the emissions reductions required in rich societies.

But we also have to consider the sources of greenhouse gases other than the burning of fossil fuels. Even if we reduce carbon fuel use to about 1,050 watts, and therefore hugely reduce carbon dioxide emissions, we will still need to deal with the much smaller but still significant emissions of methane and nitrous oxide. To be safe, we might therefore choose to completely counterbalance the carbon dioxide from 1,050 watts of fossil fuel use by carbon sequestration in soils and plants.

The last two chapters of this book, on biochar and soil improvement, demonstrate that doing so is relatively easy and probably not expensive, though it will undoubtedly be difficult to organize on a large scale. Of equal importance to the climate change benefits, storing more carbon in the soil will probably improve the agricultural productivity of the world's land, increasing the amount of food that can be harvested and reducing the pressure to cut down the world's remaining forests.

How much extra carbon do we need to store in the soil to completely offset 1,050 watts of fossil fuel power so that the total amount of carbon dioxide in the air remains the same? This calculation is quite simple and very encouraging: 1,050 watts of continuous fossil fuel use implies a total of about 9,000 kilowatt-hours a year. If this energy were all generated by burning oil, this amount would mean using about 200 gallons of oil per person, which would add about 2 tons of carbon dioxide to the atmosphere when burnt, containing somewhat less than 1,300 pounds of carbon.

The chapters on soil improvement and biochar suggest that across large swaths of the world, we can reasonably aim to increase the carbon stored in soils by at least 1 percent of the weight of the soils themselves. If this were achieved, each person's 1,050 remaining watts of continuous fossil fuel could be offset by an area of just 1,000 square feet per year. Of course, this program of carbon storage has to go on year after year until we have completely decarbonized the economy, but accomplishing that task is perfectly possible. Even the U.K., with less land per inhabitant than almost any other country in the world, could offset the emissions resulting from 1,050 watts of continuous fossil fuel use per capita for nearly thirty-five years before it ran out of space. In other parts of the world, the numbers are even more striking: Australia could sequester the carbon emissions resulting from its remaining fossil fuel use for several thousand years by achieving the same increase in soil carbon levels across the country. Increasing carbon levels in soils is also probably the cheapest way of reducing the carbon dioxide in the atmosphere.

A more intractable issue has been raised several times in this book. Sequestering carbon in the soil requires us to add plant matter, whether in the form of biochar, longer roots, or greater amounts of humus. We also need to use wood and straw for making cellulosic ethanol for fuel cells and car engines and for providing the fuel for biomass heating plants. If the guesses in this chapter are approximately correct, we may need 800 continuous watts of our total per capita energy need to come from plants and trees. This is about the same amount of energy we might get from wind or solar power in 2025.

The photosynthesis process that creates all plant matter is much less efficient at translating energy from the sun into useful energy than, for example, a solar panel of similar size. To be clear, this means that we would need far less space if we used photovoltaic panels to deliver our remaining energy needs than if we used trees to make ethanol or as fuel for a power plant. However, an acre of solar panels will always be very much more expensive than a similar area of trees. The problem is that to get 800 watts of energy from wood and plant matter will require about a third of an acre for each person in the industrial world. If a country such as France wanted to use its own land to grow the woody matter it needed, it would have to devote over 15 percent of its total area to this purpose. Of course, growing the wood or straw in less developed regions and then moving the ethanol to where it is needed would probably make more sense. This approach would have the benefit of increasing incomes in the developing world and providing a real incentive to reforest large areas. But it will still require a significant percentage of the world's usable land area to be given over to renewable forests for making the raw material for cellulosic ethanol.

It would be foolish to deny that developing a huge new industry that grows and processes billions of tons of forest matter around the world each year is a difficult challenge. Likewise, it would be foolish to think that decarbonizing the world's electric grids is a simple task. But I believe—and I hope this book has shown—that it can be done.

EPILOGUE

T HE TEN technologies, implemented and supported as a portfolio, have the capacity to tackle climate change. Few of them
can yet match conventional energy sources in terms of cost, even
at today's elevated prices of oil, coal, and gas. But we can be confident that energy produced by these new techniques will decline
in cost and will eventually be competitive with fossil fuels. Similarly, the technologies that focus on energy saving or carbon
sequestration will rapidly fall in price.

What is the basis for such optimism? Almost all manufacturing industries benefit from a learning curve. For example, at the
point at which total wind turbine production exceeded 2 gigawatts, the cost per unit of power was about 20 percent lower
than when the industry passed 1 gigawatt. This pattern of cost
reduction is characteristic of almost all complex manufacturing
and service industries. It is observed in aircraft manufacture, in
semiconductors, and even in repetitive machining operations or
clerical tasks. (One rare exception, as we'll see below, is nuclear
power plant construction.) As industries become mature, the
rate of decline in costs may fall, but the cost of almost everything
we make tends to fall by at least 10 percent every time cumula-
tive manufacturing volumes double. This will be the case, for

instance, in constructing biogas digesters on rural farms, building solar panels, or improving home insulation.

A cost decline of 10 percent may not sound like much. Nevertheless, it means that a technology growing by 40 percent a year—a rate achieved, for example, by some types of solar power, by small fuel cells, and probably by electric cars—will nearly halve in cost over the next ten years. This trend almost certainly puts onshore wind power, cellulosic ethanol, probably solar power, and possibly some marine energy technologies in a position to be cheaper than natural gas or oil as sources of energy. Carbon sequestration will also go down a steep curve of cost improvement. If governments increase the price of fossil fuels by imposing carbon taxes or tight caps on their use, this price advantage will be even more apparent. And robust support for technologies such as the application of Passivhaus techniques for existing buildings and the construction of cheap biochar kilns will also bring their price down. All of the technologies in this book, including those that involve soil improvement, are at the beginnings of their life cycle and will be made significantly cheaper in the coming decades.

Politicians and investors need to hear this point repeatedly. All the ten technologies, with the possibly exception of wind, need support today, whether in the form of carbon taxes, explicit or covert subsidy, state-sponsored research and development, or my own favorite, huge prizes for success. How should we reward, for example, the first company to successfully export a gigawatt-hour from the churning, underexploited tides of the Pentland Firth or of Vancouver Island? How about a $100 million bounty and public recognition for the CEO?

OMISSIONS FROM THE LIST OF TEN: WHAT ELSE COULD HAVE BEEN IN?

A different writer might have picked a different list of ten technologies. I want to touch briefly on some of the other candidates: nuclear power, energy efficiency, and geoengineering.

Nuclear power

With appropriate backing, nuclear power could comfortably provide most of the world's electricity needs within twenty-five years. (Its share is about 15 percent today and falling.) So why isn't nuclear power one of the ten technologies? Is this just another instance of a naive environmentalist irrationally opposing a well-understood, science-based technology in favor of untried alternatives?

I acknowledge that many of the arguments against nuclear power are weak. One such argument is that there isn't enough uranium to go around. Despite recent enormous jumps in the price of refined uranium, the cost of fuel is unlikely to ever rise much above 1 cent per kilowatt-hour of electricity output, a fraction of natural gas costs. Similarly, although current uranium fuel production levels are little changed from ten years ago, an early worldwide shortage of supply is unlikely, even if a large number of new power stations are built. The extraction costs of the mineral may increase as the producers exhaust the mines with the richest sources of uranium, but the total amount of uranium ore available worldwide will likely be sufficient for many decades of consumption. Supplies would tighten if the world added many hundreds of new reactors to the four hundred or so operating today, but uranium will probably never run out. Although it is much more widely dispersed and only available in low concentrations, uranium is more common than tin in the earth's crust, and it is even available in very dilute amounts in seawater.

Controversies over safety and the unresolved problem of how to store large quantities of nuclear waste remain. Recent leaks from reactors and processing plants in Britain and France show that although the safety record of nuclear power has been good, there is no reason for complacency. The decommissioning costs of the existing generation of nuclear power stations will be high. This issue is particularly significant in Britain, which is the only country in the world still using the earliest and rather crude designs for commercial reactors. Official estimates of the price for disposing safely of U.K. nuclear waste are rising, often by many billions of extra dollars every year. In addition, nuclear expansion almost certainly adds to the problem of the proliferation of weapons-grade radioactive material.

These particular problems may or may not be overcome, but one further issue always remains. For reasons we do not completely understand, the world is very poor at constructing nuclear power plants on time and on budget. And the problem is getting worse. Figure 4 looks at the cost of building a nuclear power station in the U.S. over the last few decades, expressed as the price the operator would have to charge to earn a reasonable return on the capital used to build the plant. The upward drift in construction costs is obvious, and some of the later nuclear power stations needed prices well in excess of 10 cents per kilowatt-hour—far more than competing technologies—to earn a reasonable return. Many experts expect companies such as First Solar to achieve a lower figure for solar panel installations by 2012.

Of critical importance, nuclear power stations have been getting more expensive as the technology matures, not less. This trend is in direct contrast to all the ten technologies chosen for inclusion in this book, which are all tending to get less costly the more that we make.

Recent years have seen relatively few new nuclear power stations. The nine built in Asia in the last fifteen years have cost an average of almost $3,000 per kilowatt-hour of capacity, if the cost is inflated to today's price level. This figure is far higher than the proponents of nuclear energy claim in their promotional materials. For example, the World Nuclear Association says the figure could be $1,500 a kilowatt-hour but provides no evidence to support this claim.

The price of building a nuclear station is the most important determinant of how much its electricity will cost to produce, so figures for construction costs are crucial to any understanding of whether nuclear power is genuinely competitive with the technologies discussed in this book.

FIGURE 4: *Construction costs for U.S. nuclear reactors.*

Data supplied by Jonathan G. Koomey, Stanford University

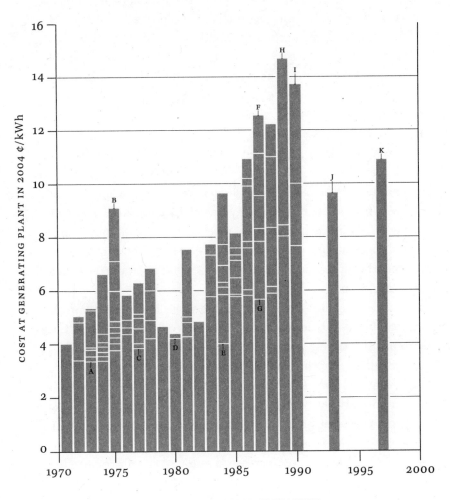

A Oconee 1

B Rancho Seco

C Browns Ferry 3

D Arkansas 2

E McGuire 2

F Clinton

G Byron 2

H Shoreham (assumes
average operation)

I Comanche Peak 1

J Comanche Peak 2

K Watts Bar 1

The most recent example of construction cost overrun is a new reactor called OL3, sited next to two existing nuclear plants at Olkiluoto in the west of Finland. Started in 2004, this project has experienced lengthy delays and may now not be ready until 2012 or later. The cost has risen substantially, possibly to double the initial estimates. The French construction company AREVA is having to pay for the overruns, and the eventual cost will likely be about $9 billion, or almost $9,000 per kilowatt-hour, three times the price of the Asian reactors. This isn't an isolated example. Some U.S. utilities are now openly forecasting that American nuclear power plants might cost as much as $8,000 or $10,000 per kilowatt-hour even though construction costs are generally lower in the U.S. Indeed, many electricity companies are shying away from nuclear because of a belief that investors will simply not fund the risk of excessively high costs. This is a clear lesson from the international capital markets that governments in favor of nuclear power should not ignore.

All types of large infrastructure projects around the world, including construction of major gas and coal power stations, have tended to become more expensive over the last few years, but the inflation in the nuclear industry has been truly remarkable. In the 1970s, nuclear plants could be built for $1,000 or $1,500 per kilowatt-hour at today's prices. Today, the figure might be six times this level, and there is no evidence that costs will eventually fall back to their earlier levels.

Are the nuclear cost increases a result of the inflationary effect of the prolonged boom of the early twenty-first century and the consequent shortage of spare capacity to design, build, and manage major projects? Or is it because of greater public concerns about safety or the impact of these plants on the local ecology? Why does the world nuclear industry not follow the pattern of the learning curve seen in almost every other manufacturing process? The most convincing explanation for the sustained cost increases is that the nuclear industry has not managed to settle on a single design and then work gradually to remove costs from this design by ironing out problems and inefficiencies. It might be said that the nuclear construction industry has never stopped building individual prototypes, all very different from each other. Nuclear construction costs are also increased by continuous enhancements of regulatory requirements and the need to spend huge sums to acquire permits to build the power stations and to meet public objections.

The nuclear industry claims that it has now settled on three basic designs for the power stations to be built in the next two decades. It ought, at least in theory, to be possible to reduce the costs of these three types of reactor to well below the extraordinarily high levels seen in Finland at the moment. We will find out fairly soon whether any optimism is justified. China and other Asian countries will build large numbers of nuclear plants in the next decade. The Chinese national energy plan foresees that about a quarter of the increase in electricity consumption in the next twelve years will be provided by nuclear power. If the country uses the AREVA design now being so expensively pioneered in Finland, it will be building over eighty separate new plants. This cohort of power stations will be the conclusive test of whether the latest generation of atomic technology can create electricity safely and at a competitive price.

The Chinese experience may well show that pessimism over nuclear construction costs is unwarranted. The published figures suggest that China expects construction bills of little more than $2,000 a kilowatt-hour, or less than a quarter of the prospective costs in Finland. If this Chinese figure turns out to be right, those Western countries that have turned against nuclear power will need to readdress their decision. But at the moment, the financial case for investing in nuclear is strikingly absent. Rather than gamble $8 billion or $12 billion on a couple of nuclear power plants, private investors and government would be far better advised to back the renewable electricity technologies and carbon capture processes covered in this book. Or they could support alternative nuclear power sources such as the Thorium molten salt reactors pushed by Kirk Sorensen and his colleagues, new "fast-breeder" technologies, or the small, "village-scale" plants being developed by Hyperion. These technologies may avoid some of the major problems that traditional approaches face because they either reduce the volume of the waste products or their radioactive danger and do not create an increased threat that nuclear material will be diverted to make explosive devices.

If reactors do actually cost $9,000 a kilowatt-hour, as implied by the Finnish experience so far, nuclear electricity is simply not competitive with the other main types of generation. Even if renewables fail to drop rapidly in price, it seems likely that, by 2020, coal-fired power stations with carbon capture will produce cheaper electricity than new nuclear plants. If we expect that to be true, there is no economic reason whatsoever for building nuclear generating plants. Since large numbers of new nuclear stations would also reduce the attractiveness of investing in competing baseload generation from renewable sources, the arguments for supporting nuclear are further weakened.

Energy-efficiency measures
Only one energy-efficiency technology gets on to the list of ten—super-insulated homes. The lack of attention paid to other power-saving techniques perhaps needs a little explanation. House insulation is the single most important energy-efficiency improvement the world can make, and it reduces both heating and air-conditioning needs. An individual in a cold country typically generates more carbon dioxide from home heating than from car travel. Other improvements, such as better refrigerators, or further improvements in lighting from the use of LEDS,

cannot hope to provide anything like the scale of benefit that better home insulation offers.

A typical poorly insulated northern European house might use 30,000 kilowatt-hours of heat a year, and the number might be larger in colder parts of North America. Among domestic appliances, the refrigerator usually provides most scope for improvement because new machines are very much better insulated than their equivalents of even ten years ago. However, replacing an old fridge with the best new model is unlikely to save more than 200 kilowatt-hours a year, less than 1 percent of the energy consumed heating the house. Simply put, improving the amount of electricity used by home appliances is an important ambition but not one that can hope to substantially reduce total energy demand.

Apart from poor housing insulation, the greatest source of waste in the consumption of energy is probably the internal combustion engine, which converts little more than a quarter of the energy in gasoline into the kinetic energy that gets us from place to place. This book proposes that replacing conventional car engines with the much more efficient use of batteries and electric motors is the most sensible way of minimizing this form of energy waste. So, strictly speaking, the chapter on electric cars also promotes an energy-saving technology.

After home heating and cars, the third most important target for efficiency improvements is the wasteful use of electricity in offices. The average office worker in a developed economy is responsible for about three times as much electricity consumption in a forty-hour work week as he or she is at home. Electronic appliances, such as computers and servers, are inefficient in their use of energy and are often kept on twenty-four hours a day. In summer, poor building design means that powerful air conditioning is needed for the entire week, partly to remove the heat created by this electronic equipment. Reducing office electricity use should be a high priority, but rather than needing any new technology, it simply calls for proper housekeeping and intelligent purchasing of low-energy appliances.

Geoengineering

If all else fails, can we avert global warming by emergency techniques to remove carbon from the atmosphere or block some of the sun's radiation reaching the earth? Some environmentalists rail against such "geoengineering" schemes, saying that they encourage the world to continue with rash and unsustainable consumption of fossil fuels. Nevertheless, rational governments and scientific institutions must carry out research into this topic. We need emergency fallbacks in case emissions reductions fail or we find that temperature increases begin to induce dangerous instability in our weather systems. If, for example, glacial and ice cap melting begins to speed up dramatically, which paleoclimatic evidence suggests is a real risk, then even rapid emissions reductions will have no measurable impact. The thermal momentum of the ice means that melting will continue even if temperatures are stabilized. A large percentage of the world's population will face disaster as sea levels rise and the summer flow of glacier-fed rivers declines sharply, causing severe water shortages for hundreds of millions of people. In these circumstances, the only appropriate action will probably be an attempt to quickly reduce global temperatures.

The simplest way to reduce temperature is to cut the total amount of the sun's radiation reaching the earth's surface. Blocking 1 or 2 percent of the solar energy that would otherwise reach us would be enough to counterbalance the effects of the greenhouse gases added to the atmosphere since the Industrial Revolution began. Two apparently viable techniques are canvassed for achieving this reduction—increasing pollution in the upper regions of the earth's atmosphere or inducing greater cloudiness in the lower regions of the air. We know from the eruption of Mount Pinatubo in 1991 that one violent volcanic eruption, blasting 20 million tons of sulfur dioxide 18 miles or more upward, would provide enough of a solar umbrella to reduce temperatures by half a degree Celsius (0.9 degrees Fahrenheit) or more. In the case of Pinatubo, the effect lasted three or four years, changing weather patterns around the world, probably enhancing the drought in the African Sahel and causing excess rainfall in the U.S. Events on the scale of Pinatubo are rare: the twentieth century only saw one or two eruptions of equivalent size.

We could mimic the effect of large volcanic events by shooting sulfur compounds into the stratosphere. This idea has distinguished adherents, such as Nobel Prize winner Paul Crutzen, but most climate scientists are horrified by its potential side-effects. It might work at restraining temperature rises, but it would increase the rate of ozone depletion, change weather systems, and increase acid rain. It would also do nothing to reduce the amount of carbon dioxide in the atmosphere, meaning that the oceans would continue to acidify as they absorbed increasing amounts of the gas. Among other effects, this acidification would destroy coral reefs and gradually wipe out sea life by killing plankton and fish.

Another way of reflecting sunlight is to create more low-level clouds. Paradoxically, wispy high-level clouds tend to keep heat in, but thick layers of cloudiness near the surface send light back into space. Probably the most plausible way of increasing low cloud cover would be to create a fine mist of salty ocean water and spray it upward. If done on a large and increasing scale, this approach would tend to increase the amount of cloudiness over the seas and help to decrease temperatures. One variant of this scheme is proposed by Stephen Salter, the inventor of one of the early devices for capturing wave energy mentioned in Chapter 2. His plan is to have hundreds of automatically controlled wind-powered boats shooting spray into the air. But, once again, even if the plan works, it doesn't reduce the bad effects of the increasing levels of carbon dioxide in the atmosphere. It simply masks more of the world's surface from the sun. Other schemes, such as shooting trillions of tiny mirrors out into space to reflect sunlight, have similar flaws.

Other geoengineering strategies aim to increase the capacity of the seas to store carbon dioxide. The amount of carbon stored at the bottom of the oceans is many times what is in the air, soil, or trees. One idea for taking more carbon dioxide to the sea floor is to seed parts of the southern oceans with tiny iron filings. The theory is that the growth of plankton is held back by a shortage of iron, an important nutrient. Plankton absorb carbon dioxide in a photosynthesis-like process that produces calcium carbonate for their skeleton-like internal structure. When plankton die, they fall to the bottom of the ocean, carrying the carbon dioxide with them in the form of the carbonate. So increasing the number of plankton could help sequester carbon.

Experiments have shown that extra iron does indeed increase the growth of plankton. And since the plankton in the oceans have a total weight greater than all the trees and plants on the earth's land, supplementing the ocean with iron is potentially extremely useful. However, excitement was tempered when scientists discovered that a very small percentage of the extra plankton actually fell to the bottom of the sea, where the carbon dioxide would be safely sequestered. What actually seemed to have happened was that bigger sea creatures simply ate more plankton and were themselves consumed by species further up the food chain. Most of the extra CO_2 absorbed by the plankton eventually seems to have returned to the air. A similar scheme that involves sucking cold, nutrient-rich, deep-level water up to the surface in order to improve the rate of photosynthesis by tiny sea creatures may suffer from a similar problem.

Schemes such as these are widely derided by climate scientists, who tend to believe that geoengineering projects simply compound the original problem rather than cure it. They correctly point out that humans cannot begin to comprehend many of the complexities of the world's weather and climate systems. To them, the idea that we could cleanly counteract the consequences of increasing greenhouse gas levels by simple techniques such as reflecting the sun's energy is hubris of the worst sort. The great climate scientist Wally Broecker says that our increasing greenhouse emissions are having an effect on climate analogous to poking an angry beast with a sharp stick. Geoengineering may compound the risks by poking the animal with a second stick. It is far better to reduce emissions or increase the rate of carbon capture in the soil, plants, and trees or by safe underground injection. (As we saw in Chapter 8, Broecker himself supports a scheme to chemically capture carbon dioxide from ambient air and sequester it underground, an approach that shouldn't have any climatic side-effects.)

But ruling out geoengineering entirely is surely a mistake. It is sensible contingency planning for a world that is only gradually waking up to the possible dangers of even modest further increases in atmospheric carbon dioxide levels. Although all geoengineering schemes will have risks, the possibility of unexpectedly rapid changes to the climate argues strongly for a sustained research effort. Even if we may never need to use these techniques, we need to understand the best ways to reflect greater amounts of sunlight and improve the ability of the oceans to take up carbon dioxide without increasing acidification.

A CARBON TAX

The first pages of this book drew a distinction between those who think that we are doomed as a result of our voracious appetite for fossil fuels and those who believe that our capacity for technological improvement will eventually allow us to reduce carbon dioxide emissions without impairing growth in prosperity. I have tried to steer a course between these two increasingly polarized camps. I believe that the evidence is strong that the ten technologies in this book can substantially reduce fossil fuel use within a decade or so. But if we are to move rapidly to a point at which these technologies can have real impact, they need to be supported. The crucial requirement is a high and increasing price on carbon, imposed either directly on the carbon content of goods and services or through well-designed schemes that cap emissions but allow permits to be traded. Both schemes mean that fossil fuels need to become more and more expensive, sending strong signals to energy users and encouraging them to use less coal, gas, and oil. In addition, a high carbon price—perhaps $50 per ton of carbon dioxide or more—would make almost all the technologies in this book competitive very soon.

The suggestion of a carbon tax of this size horrifies many free-marketeers, who claim it will impose extraordinary costs and cripple the economy. However, the maximum possible impact is about 1.5 percent of the GDP of the typical rich country, an upper limit that would only be reached if all the tax fed through into increasing energy prices but energy demand remained unchanged. Until last year, it seemed that the demand for fossil fuels was completely insensitive to their price. But 2009 saw encouraging signs that increased prices eventually affect the quantity of energy consumed. American gasoline demand is down, and British domestic electricity use has also fallen as people have responded to the market price. A high carbon tax would encourage energy conservation measures, meaning that the net impact on the economy might be very small indeed.

Sensible taxation policy needs to keep the cost of carbon-based energy high and increasing, whatever the market price of oil or coal. An intelligent government will continue adding to the price of carbon fuels while working to mitigate the impact on the poor, using such techniques as improving home insulation and subsidizing the use of low-carbon energy. All the ten technologies in this book will move forward far more quickly if innovators and entrepreneurs can be confident that the financial competitiveness of carbon-reduction technologies will be consistently supported by governments. This is true both for electricity generation from wind, solar, and marine sources and for carbon capture and storage techniques. A $50-per-ton carbon dioxide tax will help the world wean itself off fossil fuels and put in place huge programs for carbon capture.

What does $50 a ton mean in terms of costs to the consumer? Surprisingly little. Even the world's most polluting power stations, burning brown coal in old furnaces, generate about 2 pounds of carbon dioxide for every kilowatt-hour that they generate. A $50 carbon tax would therefore add 5 cents to the price of a kilowatt-hour. This amount is far from insignificant: at current U.S. retail prices, it might add almost 50 percent to the prices homeowners pay for electricity. But it is not an economic catastrophe, and most other uses of fossil fuels would see a much less significant change.

Scientists, entrepreneurs, activists, and investors around the world have made huge progress toward solving the global warming problem through advances in technology. Governments across the world simply need to help these people through intelligent and sustained support. In turn, electorates need to support those politicians who understand the need for coherent and sustained climate change programs that last several decades. This last point provides the primary reason I wrote this book. I wanted to demonstrate to the inhabitants of democratic societies that the world's climate problems are probably solvable at moderate cost. We need to vote for governments that are prepared to take the somewhat painful measures, *today,* to permanently reduce our need for fossil fuels. Politicians who argue that climate change is too expensive to solve must be rejected—urgently.

NOTES

page 24 *Sierra Club concluded turbines not a significant threat to birds*: Sierra Club, Press Release, May 19, 2006. More details at www.sierraclubmass.org/capewind. html.

page 57 *Clean power can be produced by solar thermal power plants*: DESERTEC Foundation, *Clean Power from Deserts* (The DESERTEC *WhiteBook*, 4th Edition). Bonn: DESERTEC Foundation, 2009. Available at www.desertec.org/fileadmin/downloads/DESERTEC-WhiteBook_en_small.pdf.

page 61 *International Energy Agency critical of feed-in tariff*: International Energy Agency, *Energy Policies of IEA Countries: Germany 2007 Review*. Paris: International Energy Agency/Organisation for Economic Co-operation and Development, 2007.

pages 74–75 *Department of Energy assessment of CSP*: Sandia National Laboratories and the National Renewable Energy Laboratory (working together as Sun-Lab), "Big Solutions for Big Problems... Concentrating Solar Power" Fact Sheet. Washington: U.S. Department of Energy, n.d.

page 84 *Pentland Firth waters violent and confused*: "Tide predictions from U.K. Admiralty" found at http://www.caithness.org/pentlandfirth/tides.htm. Accessed October 7, 2009.

page 84 *Major Struben on mills to collect tidal power*: Arthur Marinus Alexander Struben, *Tidal Power*. London/New York: Sir Isaac Pitman and Sons, 1921, pp. v–vi.

page 94 *Highly efficient propulsion of Thunniform-mode species*: From BioPower's website at http://www.biopowersystems.com. Accessed October 7, 2009.

pages 140–42 *"The first winter... easier to sell now"*: Telephone interview with Miriam Green, 2008.

page 172 *"Nothing else will prevent the world from exploding"*: Quoted in "Renault Bets on Electric," *The Independent*, May 14, 2008, http://www.independent.co.uk/news/business/analysis-and-features/renault-bets-on-electric-827716.html.

page 175 *Rudolf Diesel on agricultural sources of fuel*: This quotation is reported in "Historical Perspectives on Vegetable Oil–Based Diesel Fuels," Inform (report from the National Biodiesel Board), volume 12, November 2001. Available at www.biodiesel.org/resources/reportsdatabase/reports/gen/20011101_gen-346.pdf.

page 175 *Henry Ford on fuels from food*: "Ford Predicts Fuel from Vegetation," *New York Times*, September 20, 1925, p. 24.

page 183 *Vinod Khosla on 2007 being the year cellulosic ethanol will become a real prospect*: "Venture Capitalist Eyes Cellulosic Ethanol," Reuters on MSNBC, January 17, 2007.

pages 183–84 *Vinod Khosla on first commercial cellulosic ethanol plants*: Kimberly A. Strassel, "In Search of Alternatives," *Wall Street Journal*, March 24, 2008.

page 201 *David Crane on putting a price on carbon*: "We're Carboholics. Make Us Stop." *The Washington Post*, October 14, 2007. Available at http://www.washingtonpost.com/wp-dyn/content/article/2007/10/12/AR2007101202153.html.

page 214 *ccs not rocket science*: Staffan Gortz of Vattenfall speaking at a conference in Manchester, U.K., in July 2009. More details of Vattenfall's approach can be found here: http://www.guardian.co.uk/environment/video/2009/jul/13/manchester-report-staffan-gortz-ccs.

pages 229–32 *Robert Flanagan on biochar*: Personal communication with the author, 2008.

pages 246–47 *Tim Flannery on biochar*: On *Catalyst* (TV program), Australian Broadcasting Corporation. Aired August 23, 2007. Transcript at http://www.abc.net.au/catalyst/stories/s2012892.htm.

page 260 *Reid Bryson and Thomas Murray on grazing*: Reid A. Bryson, Thomas J. Murray, *Climates Of Hunger: Mankind and the World's Changing Weather.* Madison: University of Wisconsin Press, 1977.

page 289 *Figure of the cost of building a nuclear power station in the U.S.*: Data supplied by nuclear-energy expert Jonathan G. Koomey at Stanford University. Koomey's analysis of this issue can be found in J.G. Koomey and N. Hultman, "A Reactor-Level Analysis of Busbar Costs for U.S. Nuclear Plants, 1970–2005," *Energy Policy*, volume 35, issue 11 (November 2007), pp. 5630–5642.

INDEX